中等职业教育教材

化学化工专业英语

第二版

尹德胜 梁 晨 主 编
官春平 副主编

English for Chemistry and Chemical Engineering

化学工业出版社
·北京·

内容简介

《化学化工专业英语》第二版共25个单元，主要内容包括化学基础知识、化工单元操作、化工设备、石油化工、生物工程、精细化工和分析化学等。

本书对课文中出现的生词、化工技术术语、操作用语等都进行了注释；习题部分强化了学生对化工专业技术术语及操作用语的掌握；阅读理解部分都与化学化工知识和技能密切相关，且具有一定的趣味性，同时对生词都进行了注释；阅读材料均配有中文翻译，便于学生阅读及理解。

本书内容浅显实用，易于阅读和理解，适用性强，覆盖面宽，突出职业教育特色，可作为中等职业学校化学化工、生物工程及相关专业专业英语教材，也可供高职院校学生及相关专业人员参考使用。

图书在版编目（CIP）数据

化学化工专业英语/尹德胜，梁晨主编；官春平副主编.—2版.—北京：化学工业出版社，2022.8
中等职业教育教材
ISBN 978-7-122-41306-2

Ⅰ.①化… Ⅱ.①尹… ②梁… ③官… Ⅲ.①化学-英语-中等专业学校-教材 ②化学工业-英语-中等专业学校-教材 Ⅳ.①O6②TQ

中国版本图书馆CIP数据核字（2022）第072492号

责任编辑：旷英姿　刘心怡　　装帧设计：王晓宇
责任校对：赵懿桐

出版发行：化学工业出版社（北京市东城区青年湖南街13号　邮政编码100011）
印　　刷：北京云浩印刷有限责任公司
装　　订：三河市振勇印装订有限公司
850mm×1168mm　1/32　印张7½　字数125千字
2022年9月北京第2版第1次印刷

购书咨询：010-64518888　　售后服务：010-64518899
网　　址：http://www.cip.com.cn
凡购买本书，如有缺损质量问题，本社销售中心负责调换。

定　　价：21.00元

第二版前言

随着国家对职业技术教育发展力度的加大，社会对化学化工及生物工程专业技术人才素质要求的提高，那些既掌握化工与生物专业技能和知识，又能熟练运用化学化工、生物专业英语的技术人才将受到化工企业尤其是中外合资企业的欢迎。为此，我们将本书在第一版基础上进行修订再版，目的是通过本书的教学，进一步提高职业技术院校化学化工与生物专业学生专业英语水平及综合素质，使其适应当前形势需要。本书可作为职业技术院校精细化工、化学工艺、分析化学与生物工程等专业的专业素质课教材。

考虑到职业技术院校学生的培养目标和学生基础英语的水平，本书编写过程中努力体现以下特点。

① 本书所选内容尽量通俗易懂。专业英语的学习，既要求读者具有一定的基础英语水平，又要求其掌握相关的化学化工与生物专业知识与技能。职业技术院校学生学习英语的时间短，词汇量、专业知识的深度有限，因此，本书在编写过程中尽量使内容简单、易于理解，更适合中职学生学习，突出职业技术院校特色。

② 本书所选内容适应性强，覆盖面宽。选材内容涵盖了化学基础知识、化工单元操作、化工设备、石油化工、生物工程、精细化工、分析化学等。

③ 本书注重词汇与阅读材料的注释与翻译。本书对课文中出现的生词、化工技术术语、操作用语等都进行了注释；习题部分强化了学生对化工与生物工程专业技术术语及操作用语的掌握；阅读理解部分都与化学化工知识和技能密切相关，且具有一定的趣味性，同时对生词都给出了注释，便于学生阅读及理解；阅读材料都配有相应的中文翻译，目的是更好地帮助学生理解阅读内容，出发点也是考虑到学生的现有英语水平。

本书由广东轻工职业技术学院尹德胜、梁晨主编，广东轻工职业技术学院官春平副主编。尹德胜编写了Unit1～5、Unit18～20、Unit21～25石油化工与生物工程相关的13个单元的内容；梁晨编写了Unit10～14精细化工相关的5个单元的内容；官春平编写了Unit6～9化工设备相关的4个单元的内容；广东轻工职业技术学院焦创编写了Unit 15～17共3个单元的内容。全书由尹德胜统稿，由广东轻工职业技术学院陈金伟研究员主审。

本书在修订与出版过程中得到了化学工业出版社的

大力支持，借此机会特向他们表示衷心感谢。

本教材涉及内容广，编者水平有限，疏漏和不妥之处在所难免，恳请读者提出宝贵意见，以便完善。

编者

2022 年 1 月

Contents

Unit 1

Chemistry

There are different kinds of materials in our universe. Each material has its own characteristics, which is called its properties.

The two sciences, chemistry and physics, are important for the study of materials. Physics is concerned with the general properties, energy and physical changes. By contrast, chemistry is concerned with chemical properties and chemical changes. In chemical changes, materials are transformed into different materials. For example, nitrogen and hydrogen can be combined to ammonia. It is called chemical reactions.

Chemistry is very important in the use of materials. It relates to so many areas of human daily life. Chemists work in different fields of chemistry. Biochemists are interested in chemical processes in living plants and animals. Analytical chemists find ways to separate and identify chemical substances. Organic chemists study substances which contain carbon and hydrogen. Inorganic chemists study most of the other elements.

New Words

chemistry [ˈkemistri] *n*. 化学
characteristic [ˌkæriktəˈristik] *n*. 特性，特征
property [ˈprɔpəti] *n*. 性质，性能
physics [ˈfiziks] *n*. 物理学
transform [trænsˈfɔːm] *v*. 改变，转变，转化
nitrogen [ˈnaitrədʒən] *n*. 氮，氮气
hydrogen [ˈhaidrəudʒən] *n*. 氢，氢气
ammonia [ˈæməunjə] *n*. 氨
combine [kəmˈbain] *v*.（使）联合，（使）结合
living [ˈliviŋ] *adj*. 活的
analytical [ˌænəˈlitikəl] *adj*. 分析的，解析的
separate [ˈsepəreit] *adj*. 分开的，分离的 *v*. 分开，隔离
identify [aiˈdentifai] *vt*. 识别，鉴别
organic [ɔːˈgænik] *adj*. 器官的，有机的
inorganic [ˌinɔːˈgænik] *adj*. 无机的
contain [kənˈtein] *vt*. 包含，容纳

carbon [ˈkɑːbən] *n.* 碳

element [ˈelimənt] *n.* 元素，要素

Expressions and Technical Terms

be concerned with 与……有关，涉及……

physical change 物理变化

chemical property 化学性质

chemical change 化学变化

by contrast 和……比起来，对照

be transformed into 被转变成……

chemical reaction 化学反应

be combined to 化合成……，结合成……

relate to 涉及，与……有关

daily life 日常生活

chemical process 化工过程

Exercises

A. Answer the following questions.

1. What is physics concerned with?
2. What is chemistry concerned with?
3. What are biochemists interested in?
4. What are analytical chemists concerned with?

5. What do organic chemists study?

6. What do inorganic chemists study?

B. Translate the following into English.

1. 物理变化　　2. 化学变化　　3. 物理性质
4. 化学性质　　5. 化学反应　　6. 化学过程
7. 有机化学家　　8. 无机化学家　　9. 分析化学家
10. 生物化学家

C. Reading comprehension. After reading a passage, choose the best answer to each question.

When I was at university I studied very hard. But a lot of my friends did very little work. Jackson was one of them. He spent more time drinking in the students' union, than working in the library.

Once, at the end of term, we had to take an important test in chemistry. The test had a hundred questions. We just had to write 'True' or 'False'. While I was studying in my room the night before the test, Jackson was watching television. Jackson usually worried a lot the night before a test. But on that night he looked perfectly calm. Then he told me of his plan.

"It's very simple. There are a hundred questions and I have to get fifty correct to pass the test. I will take a coin into the examination room. I haven't studied a chemistry book for months, so I will just toss the coin. That way, I'm sure I will get half the questions right."

The next day, Jackson came cheerfully into the examination room. He sat tossing a coin for half an hour as he marked down his answers. Then he left, half an hour before the rest of us.

The next day, he saw the chemistry professor in the corridor. "Oh good," He said. "have you got the results of the test? What mark did I get?"

The professor looked at him and smiled.

"Ah, it's you Jackson. Just a minute."

Then he reached into his packet and took out a coin. He threw it into the air, caught it in his hand and looked at it.

"I'm terribly sorry, Jackson." He said. "you failed."

New Words and Phases

false [fɔːls] *adj*. 错误的

perfectly [ˈpəːfiktli] *adv*. 很，完全，完美地

calm [kɑːm] *v*.（使）平静，（使）镇定，平息

toss [tɔs] *v*. 投，掷

mark [mɑːk] *n*. 标志，分数，记号 *vt*. 做标记，打

分数

corridor [ˈkɔridɔː] *n*. 走廊

students' union 学生会

1. The writer of this story ( )

(A) is a university student.

(B) used to be a university student.

(C) does very little work.

2. Before the chemistry test Jackson was ( ) worried.

(A) normally (通常地)

(B) unusually (罕见地)

(C) naturally (自然地)

3. He was going to look at the coin ( ) he answered each question.

(A) after (B) before (C) because

4. Jackson expected to get ( )

(A) a good mark.

(B) his usual work.

(C) the pass mark.

5. The chemistry professor ( )

(A) thought Jackson's method was good.

(B) knew about Jackson's method.

拓展阅读

科技英语文体的主要特点

在科技英语文体中，句子的谓语动词常以被动语态的形式出现。如

1. Air and water can be converted into nitric acid.(空气和水能被转变成硝酸。)

2. It is called chemical reactions.(它被称为化学反应。)

虽然主动结构和被动结构意义相近，但被动结构使人一目了然。首先被动句并不提及人，对于一个科学家或科技工作者来说，过多的提及人不但没有必要，而且会引起含糊。其次主语是句子中非常重要的部分，把不提及人的这一部分放在句首，能引起读者的注意。在科技英语文体中谓语动词用被动语态形式可使句子简洁。科技英语中很多常用的被动语态结构在汉语中已有习惯的译法。如

It is considered that... 人们认为……

It is supposed that... 据推测，假定……

It is noticed that... 人们注意到……

It has been shown that... 已经表明……

It is reported that... 据报道……

be known as... 通常为…… 叫做……

be considered as... 被说成是……

be described as... 被描述为……

be defined as... 被定义为……

Reading Material

Important Ideas about Chemical Change

化学变化的重要概念

Making New Substances

In a chemical reaction

The starting substances, the reactants, react to give new different substances, the products. The changes which take place in the reaction are usually written as an equation.

reactants ⟶ products

means 'reacts to give' (sometimes the reaction conditions are written over the arrow)

Different kinds of chemical change

Decomposition

A single substance is broken

生成新物质

在化学反应里

起始物质，反应物，反应生成新的不同的物质，产物。反应中发生的变化通常用反应式表示。

反应物⟶产物

表示“反应生成”（有时候反应条件写在箭头上方）

不同类型的化学变化

分解反应

一种物质分裂变成两种或两种

down into two or more simpler substances.

以上更简单的物质。

Most metal carbonates decompose to give the oxide and carbon dioxide.

大多数金属碳酸盐分解生成氧化物和二氧化碳。

$$CuCO_3 \longrightarrow CuO + CO_2$$

Combination

化合反应

Two substances (usually elements) react together to make a single new compound.

两种物质（通常是元素）互相反应，生成一种新的化合物。

Mental and non-mental: aluminum and iodine react to give aluminum iodide.

金属和非金属——铝和碘反应生成碘化铝。

$$2Al + 3I_2 \longrightarrow 2AlI_3$$

Displacement

置换反应

These are reactions in which one element takes the place of another. Both metals and nonmetals can do displacements. e. g. More reactive metals can displace less reactive ones from their solutions.

反应里一种元素取代另一种元素，金属和非金属都能发生置换反应。如较活泼的金属能把较不活泼的金属从它们的溶液里置换出来。

$$Fe + CuSO_4 \longrightarrow FeSO_4 + Cu$$

Unit 2

Elements, Compounds and Mixtures

Most substances can be decomposed into two or more other substances. For example, water can be decomposed into hydrogen and oxygen. Table salt is easily decomposed into sodium and chlorine. However, an element can not be decomposed into simpler substances.

Compounds are composed of two or more elements. So they can be decomposed into simpler substances by chemical changes. A molecule is a small unit of a compound. If we divide a drop of water into smaller and smaller particle, we obtain a water molecule at last. A water molecule is composed of two hydrogen atoms and one oxygen atom. We cannot divide it if we don't destroy the molecule.

What are the characteristics of a mixture? If we mix the two elements sulfur and iron, do we have a compound? No, we have a mixture of the two elements. In fact, the iron and sulfur of the mixture can be separated by a magnet. But if the mixture is heated, the iron and sulfur combine to form iron (Ⅱ) sulfide (FeS). It contains 63.5 percent Fe and 36.5 percent S by weight. It is not attracted by a magnet.

New Words

compound [ˈkɔmpaund] *n*. 化合物
mixture [ˈmikstʃə] *n*. 混合，混合物
decompose [ˌdiːkəmˈpəuz] *v*. 分解
oxygen [ˈɔksidʒən] *n*. 氧，氧气
sodium [ˈsəudiəm] *n*. 钠
chlorine [ˈklɔːriːn] *n*. 氯，氯气
molecule [ˈmɔlikjuːl] *n*. 分子
unit [ˈjuːnit] *n*. 个体，单位
divide [diˈvaid] *v*. 分，划分，分开
particle [ˈpaːtikl] *n*. 粒子，微粒
obtain [əbˈtein] *vt*. 获得，得到
atom [ˈætəm] *n*. 原子
destroy [disˈtrɔi] *vt*. 破坏，毁坏，消灭
mix [miks] *v*. 使混合，混合
sulfur [ˈsʌlfə] *n*. 硫，硫黄

form [fɔːm] *n.* 形状 *v.* 形成，构成

iron [ˈaiən] *n.* 铁

magnet [ˈmægnit] *n.* 磁铁

attract [əˈtrækt] *vt.* 吸引，有吸引力

Expressions and Technical Terms

be decomposed into 被分解为

table salt 食盐

be composed of 由……组成

be separated 被分离

iron（Ⅱ）sulfide（FeS） 硫化亚铁

Exercises

A. Answer the following questions.

1. What can water be decomposed into?
2. What is table salt decomposed into?
3. What is a water molecule composed of?
4. how can the iron and sulfur of the mixture be separated?

B. Translate the following into English.

1. 氢气　　2. 氧气　　3. 食盐　　4. 钠

5. 氯气　　6. 分子　　7. 原子　　8. 铁

9. 硫　　　10. 硫化亚铁

C. Reading comprehension. After reading a passage, choose the best answer to each question.

Atoms are all around us. They are the bricks of which everything is made. Many millions of atoms are contained in just one grain of salt, but despite their small size they are very important. The way an object behaves depends on what kinds of atoms are in it and how they act.

For instance, you know that most solid objects melt if they get hot enough. Why is this? It is the effect of the heat on the object's atoms. When they are hot, they move faster.

Usually the atoms in an object hold together and give the object its shape. But if the object grows hot, its atoms move so fast that they break the force that usually holds them together. They move out of their usual places so that the object loses its shape. Then we say that the object is melting.

New Words and Phases

brick [brik] *n*. 砖，砖块
grain [grein] *n*. 细粒，颗粒

despite [dis'pait] *prep*. 不管，尽管，不论
object ['ɔbdʒikt] *n*. 物体
behave [bi'heiv] *v*. 举动，举止，行为
melt [melt] *v*.（使）融化，（使）熔化
effect [i'fekt] *n*. 结果，效果
hold [həuld] *n*. 把握，控制，掌握 *v*. 保持，支持
shape [ʃeip] *n*. 外形，形状，形态
break [breik] *v*. 打破
force [fɔːs] *n*. 力，力量
hold together *v*. 使结合

1. One grain of salt contains ( )
   (A) many millions of atoms.
   (B) several heated atoms.
   (C) one million atoms.
2. The way an object behaves depends on the ( )
   (A) kinds of atoms in it and how they act.
   (B) number of atoms in it.
3. Atoms in an object move ( )
   (A) at all times.
   (B) only when the object is heated.
   (C) whenever they grow hot.
4. Heating an object will affect ( )
   (A) the speed of its atoms.

(B) the shape of its atoms.

5. An object hold its shape because its atoms ( )

(A) usually hold together.

(B) move very fast.

(C) are very hot.

Reading Material

Elements, Mixtures, and Compounds

元素、混合物和化合物

Elements

- cannot be decomposed
- made of only one kind of atom
- two main kinds: metals and non-metals

元素

- 不能被分解
- 只有一种原子组成
- 分两类：金属和非金属

Mental elements physical properties

- conduct electricity
- ductile (can be bent and shaped)

金属元素 物理性质

- 导电
- 有延长性（能弯曲和成形）

Chemical properties

- form basic oxides, e. g. MgO, CuO
- form cations, e. g. Na^+, Ca^{2+}

化学性质

- 生成碱性氧化物，如 MgO、CuO
- 生成阳离子，如 Na^+、Ca^{2+}

Non-mental elements physical properties	**非金属元素 物理性质**
• insulators	• 不导电
• brittle (suddenly snap when loaded)	• 脆，易碎（当负重时，会断裂）
Chemical properties	**化学性质**
• form acidic oxides, e. g. CO_2, SO_2, NO_2	• 生成酸性氧化物，如 CO_2、SO_2、NO_2
• form anions, e. g. O^{2-}, Cl^-, Br^-	• 生成阴离子，如 O^{2-}、Cl^-、Br^-
Mixtures	**混合物**
• made by physical change	• 由物理变化而形成
• original properties still remain	• 保持组成物质的原有性质
• have variable composition	• 成分组成可以变化
• there is no energy change when mixing	• 混合时没有能量变化
• separated by physical changes	• 可用物理变化分离
Important examples of mixtures Air	**混合物的重要例子 空气**
• mixture of gases	• 混合气体
• main components	• 主要组成成分
nitrogen 78%	氮 78%
oxygen 21%	氧 21%
noble gases 0.9%	稀有气体 0.9%
carbon dioxide 0.04%	二氧化碳 0.04%
Petroleum	**石油**
• mixture of saturated hydrocarbons	• 是各种饱和碳氢化物的混合物

- composition changes from one source to another
- 产地不同，石油的成分也不同
- separated by fractional distillation
- 可以用分馏的方法分离

Compounds

化合物

- made by chemical change
- 由化学变化形成
- have new properties, different from reactants
- 跟反应物性质不同，它有新的性质
- have fixed composition and definite formula
- 有固定的组成，有一定的化学式
- there is an energy change during combination
- 在化学反应时有能量变化
- can only be separated by decomposition (a chemical change)
- 只能通过分解反应来分离（一种化学反应）

Important examples of compounds

water	H_2O
carbon dioxide	CO_2
ammonia	NH_3
sodium chloride	$NaCl$
methane	CH_4
calcium carbonate	$CaCO_3$

重要的化合物例子

水	H_2O
二氧化碳	CO_2
氨	NH_3
氯化钠	$NaCl$
甲烷	CH_4
碳酸钙	$CaCO_3$

Unit 3

Nomenclature of Inorganic Compounds

The systematic name of inorganic compounds considers the compound to be composed of two parts, one positive and one negative. The positive part is named and written first. The negative part, generally nonmetallic, is named second. Names of the elements are modified with suffixes and prefixes to identify the different types of compounds.

Examples:

Symbol(符号)	Element(元素)	Stem(词根)	Binary name(某化物)
F	fluorine	fluor-	fluoride(氟化物)
Cl	chlorine	chlor-	chloride(氯化物)
Br	bromine	brom-	bromide(溴化物)
I	iodine	iod-	iodide(碘化物)
O	oxygen	ox-	oxide(氧化物)
N	nitrogen	nitr-	nitride(氮化物)

续表

Symbol(符号)	Element(元素)	Stem(词根)	Binary name(某化物)
S	sulfur	sulf- or sulfur-	sulfide(硫化物)
P	phosphorus	phosph-	phosphide(磷化物)
C	carbon	carb-	carbide(碳化物)
H	hydrogen	hydr-	hydride(氢化物)

Binary compounds

Binary compounds contain only two different elements. Their name consists of two parts, the name of electropositive element and the name of electronegative element which is modified to end in ide.

Binary compounds containing a metal and nonmetals

Examples:

Formula(化学式)	Name(名称)
Na_2O	sodium oxide
CaC_2	calcium carbide
$AlCl_3$	aluminum chloride
NH_4F	ammonium fluoride
PbS	lead sulfide

Binary compounds containing two nonmetals

Using Latin prefix to indicate the number of atoms

in the molecule.

mono=1　　di=2　　tri=3　　tetra=4

penta=5　　hexa=6　　hepta=7　　octa=8

nona=9　　deca=10

Examples:

Formula(化学式)	Name(名称)
CO	carbon monoxide
CO_2	carbon dioxide
PCl_3	phosphorus trichloride
PCl_5	phosphorus pentachloride
N_2O_4	dinitrogen tetroxide
NO	nitrogen oxide
N_2O_3	dinitrogen trioxide
SO_2	sulfur dioxide

Acids containing no oxygen atom in the molecule

Examples:

Formula(化学式)	非含氧酸(hydro+词根+ic acid)
HF	hydrofluoric acid
HCl	hydrochloric acid
HBr	hydrobromic acid
HI	hydroiodic acid
H_2S	hydrosulfuric acid

Name of oxy-acids and oxy-compounds

Examples:

Formula(化学式)	含氧酸(词根+ic acid)	对应盐(词根+ate)
H_2SO_4	sulfuric acid	sulfate
HNO_3	nitric acid	nitrate
H_2CO_3	carbonic acid	carbonate
$HBrO_3$	bromic acid	bromate
亚含氧酸(词根+ous acid)		对应盐(词根+ite)
H_2SO_3	sulfurous acid	sulfite
HNO_2	nitrous acid	nitrite

Name of salts

Examples:

Formula(化学式)	Name of salt(盐名称)
Na_2SO_3	sodium sulfite
Na_2SO_4	sodium sulfate
KNO_2	potassium nitrite
KNO_3	potassium nitrate
$CaCO_3$	calcium carbonate
$MgBrO_3$	magnesium bromate

Name of bases

Examples:

Formula(化学式)	Name of base (hydroxide)碱名称(氢氧化物)
KOH	potassium hydroxide
$NH_3 \cdot H_2O$	ammonium hydroxide
$Ca(OH)_2$	calcium hydroxide

New Words

nomenclature [nəu'menklətʃə] *n*. 命名法

positive ['pɔzətiv] *adj*. 正的，阳性的

negative ['negətiv] *adj*. 负的，阴性的

nonmetallic ['nɔnml'tælik] *adj*. 非金属的

modify ['mɔdifai] *vt*. 更改，修改

suffix ['sʌfiks] *n*. 后缀

prefix ['pri:fiks] *n*. 前缀

fluorine ['flu(:)əri:n] *n*. 氟

binary ['bainəri] *adj*. 二元的

bromine ['brəumi:n] *n*. 溴

iodine ['aiədi:n] *n*. 碘

phosphorus ['fɔsfərəs] *n*. 磷

calcium ['kælsiəm] *n*. 钙

aluminum [ə'lju:minəm] *n*. 铝

ammonium [ə'məunjəm] *n*. 铵

lead [li:d] *n*. 铅

acid ['æsid] *n*. 酸

potassium [pə'tæsjəm] *n*. 钾

magnesium [mæg'ni:zjəm] *n*. 镁

base [beis] *n*. 碱

hydroxide [hai'drɔksaid] *n*. 氢氧化物

Expressions and Technical Terms

phosphorus trichloride 三氯化磷
phosphorus pentachloride 五氯化磷
dinitrogen tetroxide 四氧化二氮
nitrogen oxide 一氧化氮
dinitrogen trioxide 三氧化二氮
sulfur dioxide 二氧化硫
hydrofluoric acid 氢氟酸
hydrochloric acid 盐酸
hydrobromic acid 氢溴酸
hydroiodic acid 氢碘酸
hydrosulfuric acid 氢硫酸
sulfuric acid 硫酸
nitric acid 硝酸
carbonic acid 碳酸
bromic acid 溴酸
sulfurous acid 亚硫酸
nitrous acid 亚硝酸
sodium sulfite 亚硫酸钠
potassium nitrate 硝酸钾
potassium hydroxide 氢氧化钾
calcium hydroxide 氢氧化钙

Exercises

A. Translate the following into English.

H_2SO_4	NO	K_2CO_3
KOH	HCl	N_2O_3
$NaNO_3$	$NH_3 \cdot H_2O$	H_2S
CO_2	$NaNO_2$	$Ca(OH)_2$

B. Reading comprehension. After reading a passage, choose the best answer to each question.

After my last lecture, someone asked me to repeat the explanation of acids and bases. For those of you who didn't understand the first time, here it is again. The terms acid and base apply to two groups of compounds with opposing sets of characteristics. An acid is defined as a substance that releases protons, particles that have a positively electrical charge. A base is a substance that combines with these positively electrical protons. Because of these electrical charges, a strong base or acid in solution can readily conduct electricity. Another important fact that you should know is that when equal amounts of an acid and a base of the same strength are mixed, they neutralize each other. In a moment, we will go to the lab and see how neutralization works.

New Words and Phases

lecture [ˈlektʃə] *n.* 演讲
explanation [ˌeskspləˈneiʃən] *n.* 解释，解说，说明
term [tə:m] *n.* 术语
oppose [əˈpəuz] *vt.* 反对，使对立
define [diˈfain] *vt.* 定义，详细说明
release [riˈli:s] *vt.* 释放
proton [ˈprəutən] *n.* 质子
charge [tʃɑ:dʒ] *n.* 电荷
solution [səˌlju:ʃən] *n.* 溶液
readily [ˈredili] *adv.* 容易地
conduct [ˌkəndʌkt] *v.* 传导
electricity [ilekˈtrisiti] *n.* 电流，电，电学
neutralize [ˈnju:trəlaiz] *v.* 中和
apply to 将……应用于

1. Who is the speaker mostly to be? (   )
   (A) A nuclear physicist.
   (B) A chemical sales-person.
   (C) A chemical professor.

2. Whom is the speaker addressing to? (　　)

(A) Her students.

(B) Her clients.

(C) Her son.

3. What was the speaker asked to do? (　　)

(A) Do an experiment.

(B) Explain a concept again.

4. What is a proton? (　　)

(A) A solution of acid.

(B) A solution of base.

(C) A positively charged particle.

拓展阅读

科技英语构词法

1. 转化法

由一种词类转化成另一种词类，叫转化法。例如

water(*n*. 水) ⟶ water(*v*. 浇水)

charge(*n*. 电荷) ⟶ charge(*v*. 充电)

2. 派生法

通过加前、后缀构成一个新词，叫派生法。派生法是化工类科技英语中最常用的构词法。例如“烷烃”就是用前缀表示分子中碳原子数再加上“ane”作词尾构成的。如下表。

常见有机物构词规律

烷烃(尾缀 ane)	化学名称	烷烃(尾缀 ane)	化学名称
methane	甲烷	heptane	庚烷
ethane	乙烷	octane	辛烷
propane	丙烷	tridecane	十三烷
butane	丁烷	tetradecane	十四烷
pentane	戊烷	pentadecane	十五烷
hexane	己烷	hexadecane	十六烷
烷基取代基(尾缀 yl)	化学名称	烷基取代基(尾缀 yl)	化学名称
methyl	甲基	butyl	丁基
ethyl	乙基	pentyl	戊基
propyl	丙基	hexyl	己基
烯烃(尾缀 ene)	化学名称	炔烃(尾缀 yne)	化学名称
ethene	乙烯	ethyne	乙炔
propene	丙烯	propyne	丙炔
butene	丁烯	butyne	丁炔
pentene	戊烯	pentyne	戊炔
醇(alcohol)	化学名称	醚(ether)	化学名称
methyl alcohol	甲醇	methyl ether	甲醚
ethyl alcohol	乙醇	ethyl ether	乙醚
propyl alcohol	丙醇	propyl ether	丙醚

3. 合成法

由两个或更多的词合成一个词，叫合成法。有时需加连字符。

如：名词+名词	carbon steel	碳钢
	rust-resistance	防锈
形容词+名词	atomic weight	原子量
	periodic table	周期表

4. 压缩法

(1) 只取词头字母

CET	College English Test	大学英语考试
CAD	Computer Aided Design	计算机辅助设计

(2) 将单词删去一些字母

Lab	Laboratory	实验室
Corp	Corporation	股份公司
Exam	Examination	考试
Kilo	Kilogram	千克，公斤

Reading Material

Alkanes and Alkenes

Alkanes

Alkanes contain only carbon and hydrogen and only single covalent bonds. They have the maximum amount of hydrogen bonded to the carbon skeleton. So they are described as saturated. The alkanes are an example of a homologous series. A homologous series is a group of compounds with the same general formula—the alkanes have the formula C_nH_{2n+2} where n is the number of carbon atoms. Each of the alkane differs from the next by $—CH_2—$. Homologues show a gradual trend in physical properties and can have the same chemical reactions.

The chain of carbon atoms of alkanes is longer, the boiling point is higher. This is because longer chain molecules have bigger forces holding them together. But the more branched the chain the lower the boiling point, because branched chains cannot pack together so lightly.

The atoms of alkanes from butane upwards can be bonded together in more than one way. This is called isomerism. Isomers are molecules with the same molecular formula but different structural formulas.

Alkenes

Alkenes contain only carbon and hydrogen and have a double bond between two of the carbon atoms. They do not have the maximum amount of hydrogen bonded to the carbon skeleton. So they are described as unsaturated.

The alkenes have the formula C_nH_{2n} where n is the number of carbon atoms. Physical properties show a steady trend as chain length increases. Alkenes are more stable than alkanes when heated alone. They are also more reactive than alkanes when added to other substances.

烷烃和烯烃

烷烃

烷烃只含有碳和氢，只有单键。烷烃中跟碳键键合的氢达到最大数量，因此它们被形容成饱和的。烷烃是同系物的一个实例。同系物是一组有相同通式的化合物——烷烃的通式是 C_nH_{2n+2}，n 是碳原子的数目。

每一种烷烃跟相邻的烷烃相差—CH_2—。同系物在物理性质上表现出一种逐渐变化的趋势，它们可以有相同的化学反应。

烷烃的碳链越长，沸点越高，因为分子链越长，把分子聚合在一起的力就越大。支链越多，沸点越低，因为支链不能紧密地叠合在一起。

从丁烷起，烷烃里的原子可以用多种途径键合在一起，这就叫异构现象。异构现象是指有相同的分子式但结构式不同的分子。

烯烃

烯烃只含有碳和氢，在碳原子之间有一双键，跟碳链键合的氢未达到最大数量。因此，它们被形容成不饱和的。

烯烃的通式是 C_nH_{2n}，n 是碳原子的数目。随着碳链增长，物理性质表现出一种逐渐变化的趋势。当加热时，比烷烃更稳定。当加入其他物质时，比烷烃活泼。

Unit 4

Energy and Chemical Energy

Matter can have both potential and kinetic energy. Potential energy is stored-up energy. For example, water behind a dam has potential energy and it can be changed into electrical energy. Gasoline has chemical potential energy. It can be released during combustion.

All moving bodies have kinetic energy. When the water behind the dam is allowed to flow, its potential energy is changed into kinetic energy. The kinetic energy can be used to produce electricity.

The common kinds of energy are mechanical, chemical, heat, electrical and light energy.

In all chemical reactions, matter either absorbs or releases energy. Chemical reactions can be used to produce different kinds of energy. For example, heat and light energies are released from the combustion of fuels.

Conversely, energy can be used to cause chemical

reactions. For example, a chemical reaction occurs when light energy is used by plants. A chemical reaction also occurs when electrical energy causes water to decompose.

One type of energy may be changed into energy of another type. For example, when hydrogen and oxygen are burned, chemical energy is changed into light and heat energy. When electrical energy decomposes water again producing hydrogen and oxygen, electrical energy is changed into chemical energy.

New Words

potential [pəˈtenʃəl] *adj*. 潜在的，可能的
kinetic [kaiˈnetik] *adj*. 运动的，动的
dam [dæm] *n*. 水坝
gasoline [ˈgæsəliːn] *n*. 汽油
combustion [kəmˈbʌstʃən] *n*. 燃烧
absorb [əbˈsɔːb] *vt*. 吸收，吸引
fuel [fjuəl] *n*. 燃料
conversely [ˈkɔnvəːsli] *adv*. 相反地
cause [kɔːz] *n*. 原因 *vt*. 引起
occur [əˈkəː] *vi*. 发生，出现

Expressions and Technical Terms

potential energy　势能

kinetic energy　动能

stored-up　储存

be changed into　被转变成……

electrical energy　电能

mechanical energy　机械能

chemical energy　化学能

heat energy　热能

light energy　光能

Exercises

A. Fill in the blanks.

1. Matter can have both ________ and ________ energy.

2. All moving bodies have ________.

3. The common kinds of energy are ________ energy.

4. In all chemical reactions, matter either ________ or ________ energy.

5. Energy can be used to cause ________.

6. One type of energy may be ________ energy of

another type.

B. Translate the following into English.

1. 动能 2. 势能 3. 汽油 4. 电能
5. 机械能 6. 化学能 7. 热能 8. 光能
9. 吸收能量 10. 释放能量

C. Reading comprehension. After reading a passage, choose the best answer to each question.

In 1960, the salt density of Great Salt Lake ranged between twenty and twenty-five percent. The Dead Sea is saltier than Great Salt Lake, and both are saltier than the oceans.

In cold weather, some of the salts precipitate out since salts saturation point lowers with the temperature. Lots of almost pure sodium sulfate have piled up. The water is about one-fourth salt by weight. Imagine taking two quarts of water, boiling it, and obtaining a pound of coarse salt. Commercial salt companies along the shore take tons of salt which are refined in the basin every year. Yet an estimated six and one-half billion tons of salt remain in the water-enough to supply the world's needs for generations.

New Words and Phases

density ['densiti] *n*. 密度，浓度

range [reindʒ] *v*. 在……范围

precipitate [pri'sipiteit] *v*. 沉淀

lower ['ləuə] *v*. 降低，减弱

imagine [i'mædʒin] *vt*. 想象，设想

quart [kwɔːt] *n*. 夸脱（容量单位）

boil [bɔil] *v*. 煮沸

coarse [kɔːs] *adj*. 粗糙的

refine [ri'fain] *vt*. 精炼，精制

basin ['beisn] *n*. 盆，盆地，水池

estimate ['estimeit] *v*. 估计，估价，评估

remain [ri'mein] *vi*. 保持，剩余

generation [ˌdʒenə'reiʃən] *n*. 产生

saturation point *n*. 饱和点

sodium sulfate *n*. 硫酸钠

1. The salt density of Great Salt Lake is (　　)

(A) less than that of the Pacific Ocean.

(B) greater than that of the Pacific Ocean.

(C) greater than that of the Dead Sea.

2. Sodium sulfate is (　　)

(A) a salt.　　(B) a sand.　　(C) part silt.

3. Boiling two quarts of Salt Lake water produces (　　)

(A) one pound of salt.

(B) one quart of salt.

(C) two quarts of salt.

4. The salt on your table is (　　)

(A) coarse salt.

(B) refined before it is sold.

5. The water of Great Salt Lake contains about (　　)

(A) 6.5 million tons of salt.

(B) 6.5 tons of salt.

(C) 6.5 billion tons of salt.

拓展阅读

科技英语文体的主要特点

1. 科技人员趋于使用一些显得稳重的规范词（formal words）（因为普通词汇一般来说显得比较随便），这样从词汇这一面突出了科技英语正式、庄重的语体特征，避免使用所出现的较正式的词语。下面列出一些普通英语词和科技英语中的非技术词。

普通英语词（一般语体）	科技英语中的非技术词（正式语体）	词　义
about	approximately	大约
ask	inquire	询问
begin	commence	开始，着手
buy	purchase	买，购买
change	transform	转换，改变
cheap	inexpensive	便宜的
finish	complete	完成
get	obtain	获得，得到
give	accord	给予
have	possess	占有，拥有
method	technique	技术，方法
quick	rapid	迅速的
try	endeavor	努力，尽力
use	employ	使用
fire	flame	火焰
happy	excited	兴奋的
careful	caution	小心
heart	center	中心，中央
enough	sufficient	充分的，足够的
in the end	eventually	最后，终于

2. 一些短语动词往往由正式动词代替，表现出科技英语行文要求精练，表达上力避繁冗。例如

短语动词	正式动词	词义
take in	absorb	吸收，吸引
push in	insert	插入
put up	erect	使直立，树立
put out	extinguish	熄灭
wear away	erode	侵蚀，腐蚀
take away	remove	移动，迁移
use up	consume	消耗，消费
carry out	perform	履行，执行
come across	encounter	遭遇，遇到

Reading Material

Matter and Energy

Incomplete combustion

- In the open air, a fuel reacts with oxygen until it is used up. The fuel is completely combusted.
- If the supply of air is limited, the oxygen in the air may be used up before the fuel. The fuel is incompletely combusted.
- If a hydrocarbon is incompletely

物质和能

不完全燃烧

- 在敞开的空气中，燃料跟氧气反应，直到燃料用完，这时燃料完全燃烧。
- 如果空气是有限的，空气里氧气就可能在燃料烧完以前就已经耗尽，这时燃料是不完全燃烧。
- 如果一种烃是不完全燃烧，烃

combusted, the carbon in it may be:

里的碳可能：

only partially oxidized, forming monoxide.

只有部分氧化，生成一氧化碳。

unburned, forming soot.

燃烧不充分，生成煤烟。

- In car engines, turbos pump more air into the engine. This improves efficiency because it helps the fuel to combust completely.

- 在汽车发动机里，涡轮叶片把更多的空气打入发动机里，以提高效率，因为它使燃料充分燃烧。

The interaction of matter and energy

物质和能量的相互作用

- In chemistry we study the effects of applying energy to matter or getting matter to release stored energy.

- 在化学里，我们学习能量对物质的作用以及通过释放储存的能量来获取物质。

- In chemical plants, energy, heat, and pressure are applied to raw materials like crude oil or iron ore, changing them into useful substances that we need.

- 在化工厂里，能量、热和压力常用来加工原材料，比如原油和铁矿石，把它们变成我们需要的有用物质。

- In power stations, fuels react with air. The potential energy stored in the fuel is released as kinetic (heat) energy which is made to do work generating electricity.

- 在发电厂里，燃料跟空气反应，储藏在燃料里的势能释放出来变成动（热）能，它们主要用来做功发电。

Unit 5

Catalysis

A catalyst is very important in chemical reactions. It can increase the rate of a reaction, but it is not used up itself in the reaction. For example, in the hydrolysis of an ester, hydroxide is not required in the reaction, but its addition increase the reaction rate.

$$CH_3COOC_2H_5 + H_2O \xrightleftharpoons{OH^-} CH_3COOH + C_2H_5OH$$

We can call the hydroxide ion a catalyst.

Now how do catalysts increase reaction rate? We can use activation energy to explain. If the activation energy of a reaction is high, the reaction rate is low. But the activation energy can be lowered by catalysts, so the rate of a reaction is increased. Note that the catalyst has the same effect on the reverse reaction. So if a catalyst doubles the rate in the forward reaction, it also doubles the rate in the reverse reaction.

New Words

catalysis [kəˈtælisis] *n.* 催化作用

catalyst [ˈkætəlist] *n.* 催化剂

increase [inˈkriːs] *n.* 增加 *v.* 增加

hydrolysis [haiˈdrɔlisis] *n.* 水解

ester [ˈestə] *n.* 酯

ion [ˈaiən] *n.* 离子

explain [iksˈplein] *v.* 解释，说明

note [nəut] *vt.* 注意

double [ˈdʌbl] *vt.* 使加倍

Expressions and Technical Terms

the rate of a reaction 反应速率

used up 用完，消耗完

activation energy 活化能

reverse reaction 逆反应

forward reaction 正反应

Exercises

A. Answer the following questions.

1. What can a catalyst do in a chemical reaction?

2. Can the hydroxide ion be a catalyst in the hydrolysis of an ester?

3. Can the activation energy be lowered by catalysts?

4. Does the catalyst have the same effect on the reverse reaction?

B. Translate the following into English.

1. 催化作用	2. 催化剂	3. 反应速率
4. 活化能	5. 水解	6. 酯
7. 氢氧化物	8. 正反应	9. 逆反应

C. Reading comprehension. After reading a passage, choose the best answer to each question.

There are three main groups of oils: animal, vegetable and mineral. Great quantities of animal oil come from whales, those enormous creatures of the sea which are the largest remaining animals in the world. To protect the whale from the cold of the Arctic seas, nature has provided it with a thick covering of fat called blubber. When the whale is killed, the blubber is stripped off either on board or on shore. It produces a great quantity of oil which can be made into food for human consumption. A few other creatures yield oil, but none so much as the whale.

Vegetable oil has been known from antiquity. No household can get on without it, for it is used in cooking. Perfumes may be made from the oils of certain flowers. Soaps are made from vegetable and animal oils.

To the ordinary man, one kind of oil may be as important as another. But when the politician or the engineer refers to oil, he almost always means mineral oil, the oil that drives tanks, aeroplanes and warships, motor-cars and diesel locomotives; the oil that is used to lubricate all kinds of machinery. This is the oil that has changed the life of the common man. When it is refined into petrol it is used to drive the internal combustion engine. To it we owe the existence of the motor-car, which has replaced the private carriage drawn by the horse. To it we owe the possibility of flying. It has changed the methods of warfare on land and sea. This kind of oil comes out of the earth. Because it burns well, it is used as fuel and in some ways it is superior to coal. Many big ships now burn oil instead of coal. Because it burns brightly, it is used for illumination.

Countless homes are still illuminated with oil-burning lamps. Because it is very slippery, it is used as lubrication. Two metal surface rubbing together cause friction and heat. But if they are separated by a thin film of oil, the friction and heat are reduced. No machine would work for long if it were not properly lubricated. The oil used for this purpose must be of the correct thickness. If it is too thin it will not give sufficient lubrication, and if it is too thick it will not reach all parts that must be lubricated.

New Words and Phases

vegetable [ˈvedʒitəbl] *n*. 蔬菜，植物
mineral [ˈminərəl] *n*. 矿物，矿石
whale [weil] *n*. 鲸
creature [ˈkriːtʃə] *n*. 动物
provide [prəˈvaid] *v*. 供应，供给
blubber [ˈblʌbə] *n*. 鲸脂
strip [strip] *n*. 条，带
quantity [ˈkwɔntiti] *n*. 量，数量
consumption [kənˈsʌmpʃən] *n*. 消费
yield [jiːld] *n*. 产量，收益

antiquity [æn'tikwiti] *n*. 古代，古老，古代的遗物

perfume ['pəːfjuːm] *n*. 香味，芳香，香水

tank [tæŋk] *n*. 槽，箱，罐，釜

aeroplane ['ɛərəplein] *n*. 飞机

warship ['wɔːʃip] *n*. 军舰，战船

diesel ['diːzəl] *n*. 柴油机

locomotive [ˌləukə'məutiv] *n*. 机车，火车头

lubricate ['luːbrikeit] *vt*. 润滑

petrol ['petrəl] *n*. 汽油

owe [əu] *v*. 把……归功于，欠

existence [ig'zistəns] *n*. 存在，实在

carriage ['kæridʒ] *n*. 马车，客车

warfare ['wɔːfɛə] *n*. 战争

superior [sjuː'piəriə] *adj*. 较高的，上好的

illumination [iˌljuːmi'neiʃən] *n*. 照明，阐明，启发

countless ['kautlis] *adj*. 无数的，数不尽的

slippery ['slipəri] *adj*. 滑的，光滑的

friction ['frikʃən] *n*. 摩擦，摩擦力

strip off *v*. 剥落

internal combustion engine 内燃机

1. Whales are (　　)

(A) the largest animals living on land.

(B) the largest animals now living in the

world.

2. Vegetable oil (　　)

(A) as unknown in ancient times.

(B) as known to people long ago.

(C) known only to old people.

3. The term "mineral oil" is used by the author to refer to (　　)

(A) the oil from which petrol is made.

(B) petrol only.

(C) any oil that burns.

4. The purpose of lubrication is (　　)

(A) to produce heat.

(B) to reduce friction.

Reading Material

Factors Affecting Reaction Rates

影响反应速率的因素

Concentration

- Increasing the concentration increases the number of particles.
- Increasing the number of particles increases the number of col-

浓度

- 浓度增加，反应物的粒子数增加。
- 粒子数增加则粒子之间的碰撞数增加。

lisions.

- Increasing the number of collisions increases the number of successful collisions.

This increases the rate of reaction.

- 碰撞次数增加，则有限碰撞次数也增加。

这样反应速率就加大。

Pressure

Increasing the press means that the gas molecules are squashed into a smaller volume. The same amount of gas in a smaller volume has a greater concentration. There will be more collisions, so there will be more successful collisions, so the rate will increase.

压力

压力增加表示气体分子被挤压进一个更小的空间。在一个较小的空间，同样数量的气体浓度就大了。这样碰撞更多，有效碰撞也更多，反应速率加大。

Temperature

Not all collisions between reactants succeed in making products. Only those collisions with enough energy to break bonds in the reactants will lead to a reaction. The energy is called the activation energy. Increasing the temperature of the reaction means more particles have the activation energy.

温度

并不是反应物之间所有的碰撞都能导致生成产物，只有那些有足够能量以打破反应物内的化学键的碰撞才能引起反应。这个能量叫做活化能。增加反应温度表示有更多的粒子具有活化能。

Catalyst

A catalyst allows the reaction to go by a different pathway with a lower activation energy. More particles will have this lower activation energy, and so more collisions will be successful. More successful collisions means a higher rate. e. g. iron is added as catalyst in the Harber process for making ammonia.

催化剂

催化剂能改变反应的途径，这种途径只要较低的活化能。由于更多的粒子具备了这种较低的活化能，这样，更多的碰撞就会是有效的。更多的有效碰撞表示更高的反应速率。比如，铁作为哈伯法中生产氨的一种催化剂。

Unit 6

Solubility, Solutions and Suspensions

If some sugar is added to water and stirred, the sugar disappears. In other words, it dissolves, because sugar is soluble in water. It is said to have formed a solution and sugar is the solute. The solubility of a substance is defined as the mass of that substance which will dissolve in 100g of water at a defined temperature. For example, when 104g of potassium nitrate is dissolved in 100g of water at 60℃, a saturated solution is formed. For most solids, the solubility increases with increasing temperature. For gases, however, the solubility decreases with increasing temperature.

If some flour is added to water and stirred, the flour does not disappear. In other words, it does not dissolve, because flour is insoluble in water. Instead of dissolving, the particles of flour are suspended in the water. A suspension has been formed.

New Words

solubility [ˌsɔljuˈbiliti] *n*. 溶解度，溶解性
suspension [səsˈpenʃən] *n*. 悬浮液，悬浮
sugar [ˈʃugə] *n*. 糖
stir [stəː] *vt*. 摇动，搅和
dissolve [diˈzɔlv] *v*. 溶解
soluble [ˈsɔljubl] *adj*. 可溶的，可溶解的
solute [ˈsɔljuːt] *n*. 溶质
mass [mæs] *n*. 块，质量
saturate [ˈsætʃəreit] *v*. 饱和
flour [ˈflauə] *n*. 面粉
insoluble [inˈsɔljubl] *adj*. 不能溶解的
suspend [səsˈpend] *vt*. 悬挂，悬浮

Expressions and Technical Terms

in other words 换句话说
potassium nitrate 硝酸钾
saturated solution 饱和溶液
Instead of 代替，而不是

Exercises

A. Answer the following questions.

1. What happens when you add some sugar to water and stir it?

2. What is the definition of solubility?

3. What is formed when 104g of potassium nitrate is dissolved in 100g of water at 60℃?

4. What happens when you add some flour to water and stir it?

B. Translate the following into English.

1. 溶解度　　2. 溶液　　3. 悬浮液

4. 溶解　　5. 硝酸钾　　6. 溶质

C. Reading comprehension. After reading a passage, choose the best answer to each question.

A suspension is a dispersion of microscopic particles in a fluid. When you add some flour to water and stir it, a suspension will be formed. Gradually, the particles of flour begin to settle, until they are deposited at the bottom of the water, which becomes clear again. The smaller the particles, the slower they settle; the larger the particles, the quicker they settle.

The particles in a suspension are visible with the aid of a microscope. But if the suspended particles are so small that they would never settle, and are invisible even with the aid of a microscope. They are said to have formed a colloid. A colloid is a suspension of submicroscopic particles.

A suspension differs from a colloid in the size of the particles. Colloidal particles are charged with electricity, most of them negatively.

New Words and Phases

dispersion [dis'pəːʃən] *n*. 散布，驱散，传播
microscopic [maikrə'skɔpik] *adj*. 用显微镜可见的
fluid ['fluːid] *n*. 流体，流动性
settle ['setl] *v*. 澄清，沉淀
deposit [di'pɔzit] *n*. 堆积物，沉淀物，*v*. 存放，堆积
bottom ['bɔtəm] *n*. 底，底部
microscope ['maikrəskəʊp] *n*. 显微镜
visible ['vizəbl] *adj*. 看得见的
colloid ['kɔlɔid] *n*. 胶体
differ ['difə] *vi*. 不一致，不同

1. The particles in a suspension ( )

(A) would not deposit at the bottom of the water.

(B) are visible with the aid of a microscope.

(C) are charged with electricity.

2. The particles in a colloid (　　)

(A) would not settle at the bottom of the water.

(B) are visible with the aid of a microscope.

(C) are larger than the particles in a suspension.

Reading Material

Hard Water

Water containing dissolved calcium or magnesium ions is said to be hard. In hard water, soap does not lather as well and forms a scum. The calcium and magnesium ions get into the water in two ways. Firstly, rain falls on soluble calcium or magnesium minerals and dissolves them. Secondly, rain (which is slightly acidic) reacts with basic calcium or magnesium minerals. Rainwater is acidic because it contains dissolved carbon dioxide. So rain is dilute carbonic acid.

The advantages of hard water are better taste, better for brewing beer and healthier (less heart disease, stronger bones and teeth). But its disadvantages are that it can form a scum with soap and forms a scale in hot water boilers and kettles.

Water which has had its hardness removed is described as softened water. Water is softened in two ways. Firstly, by adding sodium carbonate the calcium and magnesium ions are precipitated out. Secondly, the hard water is run through a resin which replaces the calcium or magnesium ions with sodium ions by using ion exchangers.

硬水

水中含有溶解的钙或镁离子，就称这种水变硬。在硬水里，肥皂不易发泡、去污。钙、镁离子通过两种途径进入水中。第一种，雨水降落在含钙或镁的矿物中，矿物溶解。第二种，雨水（稍带酸性）跟碱性的钙、镁矿反应。雨水因为含有溶解的二氧化碳而呈酸性，因此可以说雨水是很稀的碳酸。

硬水的优点是口感好，有利于酿造啤酒，有利健康（减少心脏发病，强化骨骼和牙齿）。硬水的缺点是跟肥皂形成浮渣，在水壶里和锅炉里生成水垢。

使水去掉硬度叫水的软化。水的软化有两种方法。首先，加入碳酸钠生成沉淀，从而去掉钙、镁离子。其次，也可使用离子交换剂，硬水流过一种树脂，树脂里的钠离子会取代钙、镁离子。

Unit 7

Heat Transfer

Heat always tends to pass from warmer substances to cooler substances. When a warm substance contacts with a cold substance, the molecules of the warm substance give its energy to the cold molecules.

In a chemical plant, transfer of heat is very important. Three methods of heat transfer are conduction, convection and radiation.

Conduction occurs when we place a spoon in a cup of hot coffee. The spoon will get hot. Heat is passed from the coffee to the spoon. The spoon becomes hot and the coffee becomes a little cooler.

Convection is the transfer of heat from one place to another within a liquid or gas. When a gas or liquid is heated, it becomes lighter and moves upward. Then the heavier gas and liquid comes there.

When heat is transferred by heat waves, it is called radiation. The heat from the sun is an example of radiation.

New Words

transfer [træns'fəː] *n*. 迁移，移动 *v*. 转移，传递

conduction [kən'dʌkʃən] *n*. 传导

convection [kən'vekʃən] *n*. 对流

radiation [ˌreidi'eiʃən] *n*. 辐射，发散，发光

spoon [spuːn] *n*. 匙，勺子

Expressions and Technical Terms

heat transfer 传热

pass from 传递

contact with 与……接触

chemical plant 化学工厂

Exercises

A. Fill in the blanks.

1. Heat always tends to pass from ________ substances to ________ substances.

2. Three methods of heat transfer are ________, ________ and ________.

3. ________is the transfer of heat from one place to another within a liquid or gas.

4. When heat is transferred by ________, it is called radiation.

B. Translate the following into English.

1. 化工厂　　2. 传热　　3. 传导

4. 对流　　5. 辐射

C. Reading comprehension. After reading a passage, choose the best answer to each question.

The existence of oil wells has been known for a long time. Some of the Indians of North American used to collect and sell the oil from the wells. No one, however, seems to have realized the importance of this oil until it was found that paraffin-oil could be made from it. This led to the development of the wells and to the making of enormous profits. When the internal combustion engine was invented, oil became of world-wide importance.

What was the origin of the oil? Scientists are confident about the formation of coal, but they do not sure when about oil. They think that the oil under the surface of the earth originated in the distant past, and was formed from living things in the sea. Countless billions of sea creatures and plants lived and sank to the sea bed. They were covered with huge deposits of mud. By processes of chemistry, pressure, temperature and through long ages they were changed to oil. In some places gas and oil come up to the surface of sea bed. Very few of them are far distance from the oceans of today. The rocks in which oil is found are of marine origin too.

New Words and Phases

well [wel] *n*. 井

realize ['riəlaiz] *vt*. 认识到，了解

paraffin-oil ['pærəfin-ɔil] *n*. 石蜡油

profit ['prɔfit] *n*. 利润，益处，得益

confident ['kɔnfidənt] *adj*. 自信的，确信的

pressure ['preʃə(r)] *n*. 压，压力

marine [mə'riːn] *adj*. 海的，航海的

1. Mineral oil became world-wide important (   )

   (A) the internal combustion engine was invented.

   (B) American Indians began to collect and sell it.

2. Scientists think that ( )
   (A) oil was formed from sea creatures.
   (B) oil was formed from sea water.

Reading Material

Heat Transfer and Heat Exchangers

传热和换热器

The importance of heat transfer

- In the major of chemical processes, heat is either given out or absorbed, and in very range of chemical plant, fluids must be either heated or cooled. Thus in furnaces, evaporators, distillation units, driers, and reactors, one of the major problems is that of transferring heat at the desired rate.
- Alternatively, it may be necessary to prevent the loss of heat from a hot vessel or steam pipe. The control of the flow of heat forms one of the most important

传热的重要性

- 在大部分化工过程中，热量或被释放出或被吸收。在许多化工厂中，流体常需被加热或被冷却。如加热炉、蒸发器、蒸馏单元、干燥器和反应器中一个主要的问题就是以期望的速度传热。
- 同样，避免热容器或蒸汽管道里的热量损失也是必要的。控制热量的传递成了化学工程中一个最重要的部分。

sections of chemical engineering.

The driving force of heat transfer

- In order for heat to flow, there must be a driving force. This driving force is the temperature difference between the points where heat is received and where the heat originates.
- In the study of the flow of fluids, it has been observed that a fluid tends to flow from a point of high pressure to one of low pressure. The driving force, in this case, is the pressure drop. Similarly, heat tends to flow from a point of high temperature to a point of low temperature because of temperature-difference driving force.

热量传递的驱动力

- 为了让热量传递，一定要有一个驱动力。这个驱动力就是热量接受和热量产生这两点之间的温度差。
- 在流体流动的研究中，人们已观察到流体趋向于从高压处流向低压处。这种情况下的驱动力是压力降。同样，由于温度差这一驱动力，热量趋向于从高温处流向低温处。

Classification of heat exchangers

- Heat transfer equipment (heat exchangers) may be defined as apparatus in which heat is transmitted from one fluid to another. According the different methods of carrying out the heat transfer operations, heat exchangers may be classified into

换热器的分类

- 传热装置（热交换器）可定义为将热从一流体传递给另一流体的设备。根据热传递操作方法的不同，热交换器可分成以下三种基本类型：

the following three basic types:

1. direct contact heat exchangers
2. Recuperators
3. Regenerators

1. 接触型热交换器
2. 同流换热器
3. 交流换热器

Recuperators type

- The heat exchangers used in most chemical plant are of the recuperators type. In this type of equipment, heat is transferred by convection from the solid wall, and by conduction through the wall.
- The many types of recuperative heat exchangers may be classified into a number of categories, as follows:

1. By the function the heat exchangers fulfill in a process: (1) heaters, (2) coolers, (3) condensers.

2. By the kind of working media and their state of aggregation: (1) vapor-liquid heat exchangers, (2) liquid-liquid heat exchangers, (3) gas-liquid heat exchangers, (4) gas-gas heat exchangers.

同流换热器

- 在大部分化工厂里使用的是同流换热器。这种类型的换热器中，在固体壁是通过对流传热，穿过壁时通过传导传热。
- 多种同流换热器可以按以下许多类型进行分类：

1. 按过程中热交换器行使的功能来分类：(1) 加热器，(2) 冷却器，(3) 冷凝器。

2. 按工作介质的种类及它们组合的状态来分类：(1) 蒸汽-液体热交换器，(2) 液体-液体热交换器，(3) 气-液热交换器，(4) 气-气热交换器。

Unit 8

Chemical Manufacturing Process

Typical chemical processes are shown in following figure.

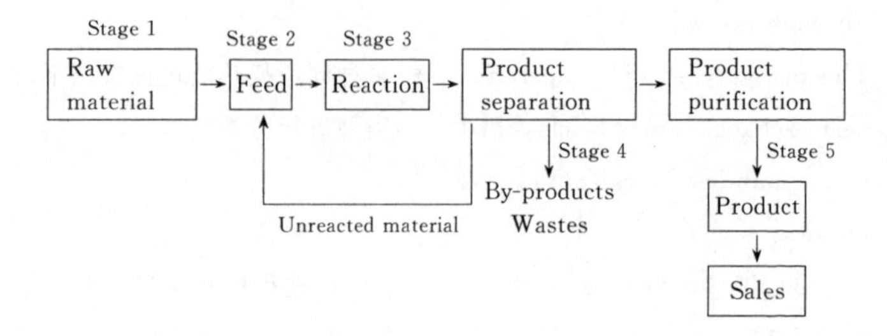

Stage 1. Raw material

The raw materials are often transported by ship, by road and rail. They are made to hold a few hours or a few days storage.

Stage 2. Feed

Some raw materials must be pure or in the right form before they are fed to the reactor. For example, some liquid feeds will need to be vaporized. Solids may need crushing, grinding and screening.

Stage 3. Reaction

The reaction stage is the heart of a chemical manufacturing process. In the reactor, the products will be formed and some by-products and unwanted compounds will also be formed.

Stage 4. Product separation

In this stage, the wastes and by-products are separated out. The unreacted material will be recycled to the reactor.

Stage 5. Product purification

Before sale the main product will usually need purification. Some by-products may also be purified for sale.

New Words

manufacture [ˌmænjuˈfæktʃə] *vt*. 制造，加工

figure [ˈfigə] *n*. 图形

stage [steidʒ] *n*. 阶段，时期

transport [trænsˈpɔːt] *vt*. 传送，运输

storage [ˈstɔridʒ] *n*. 储藏，存储

feed [fiːd] *v*. 进料，加料

separation [sepə'reiʃən] *n*. 分离，分开
reactor [ri(:)'æktə] *n*. 反应器
vaporize ['veipəraiz] *v*.（使）蒸发
crush [krʌʃ] *vt*. 压碎，碾碎
grind [graind] *v*. 磨碎，碾碎
screen [skri:n] *v*. 分筛
purification [ˌpjuərifi'keiʃən] *n*. 净化，纯化
purify ['pjuərifai] *v*. 净化，纯化

Expressions and Technical Terms

chemical manufacturing process 化学加工过程
raw material 原材料
be fed to 被填加入……
by-product 副产品
be separated out 被分离出……
be recycled to 被循环到……

Exercises

A. Fill in the blanks.

1. The raw materials are made to hold a few hours or a few days ________.

2. Some solid raw materials may need and ______,

______ and ______.

3. In the reactor, ______ will be formed and some ______ and ________ will also be formed.

4. In the product separation, the unreacted material will be ________ to the reactor.

5. Before sale the main product will usually need ________.

B. Translate the following into English.

1. 化学生产过程　2. 原材料　3. 副产品

4. 产品分离　5. 产品纯化　6. 未反应的物质

C. Reading comprehension. After reading a passage, choose the best answer to each question.

The ordinary form of mercury thermometer is used for temperatures ranging from −40℉ to 500℉. For measuring temperatures below −40℉, thermometers filled with alcohol are used. These are, however, not satisfactory for use at high temperatures. When a mercury thermometer is used for temperatures above 500℉, the space above the mercury is filled with some inert gas, usually nitrogen or carbon dioxide. As the mercury rises, the gas pressure is increased. So that it is possible to use these thermometers for temperatures exceeding 1000℉. This is the limit, however, as the melting point of glass is comparatively low. For temperatures exceeding 800℉, some form of pyrometer is generally used. The simplest of these is the mechanical pyrometer.

New Words

mercury [ˈməːkjuri] *n*. 水银，汞
thermometer [θəˈmɔmitər] *n*. 温度计，体温计
measure [ˈmeʒə] *vt*. 测量
alcohol [ˈælkəhɔl] *n*. 酒精，酒
space [speis] *n*. 空间
inert [iˈnəːt] *adj*. 惰性的
exceed [ikˈsiːd] *vt*. 超越，胜过
comparatively [kəmˈpærətivli] *adv*. 比较地，相当地
pyrometer [ˌpaiəˈrɔmitə] *n*. 高温计

1. What topics is described in this passage? (　　)
   (A) mercury thermometers.
   (B) measuring temperatures.
   (C) temperature range.

2. when we measure temperatures of about 300℉, which thermometer we choose? (　　)

(A) pyrometer.

(B) mercury thermometer.

(C) inert gas thermometer.

(D) alcohol thermometer.

Reading Material

Process Design

A process design is a network of interconnected vessels joined by pipes. Obviously it is important to calculate the flows through every vessel and pipe so that they can be correctly sized. This must be done not only for the conditions, which exist during normal production but also for those conditions, which appertain during the start-up period and for those conditions which could occur in the event of an emergency shut-down. Certain pipes and vessels are specified solely for the latter purpose. Although it is hoped that it will not be necessary to use the emergency systems, periodic testing of it (particularly the associated controls) is essential.

Knowledge of the mass and volumetric flow rates for each part of the plant results from solution of the material balance. Energy balances are also important since questions such as "what is the thermal duty?" "how much steam is required?" and "will we have a sufficient cooling water?" need to be answered.

For a process plant consisting of a mixer, reactor and separator, the balances can be performed unit by unit. The product stream from the mixer is one of the feed streams to the reactor, while the product stream from the reactor is the feed to the separator. However, sequential calculations are impossible if a flow stream from a downstream unit is returned to a unit upstream.

Recycle problems arise whenever there is a need to recycle unconverted reactant. Even near equilibrium conversion in the Haber process for ammonia would achieve only 25 percent conversion and so the nitrogen and hydrogen are recycled. The overall conversion of reactants into products can be made to be nearly 100 percent. The synthesis of methanol from carbon monoxide and hydrogen is similar in that a recycle is essential because the conversion is again about 25 percent. The production of vinyl chloride from ethylene also involves recycle streams, and many more examples of this sort could be given. More complex processes often involve several recycle loops, and at the process design stage consideration has to be given to the effect they will have not only on steady-state operation but also on the start-up process.

Although the type of work outlined above will have been unfamiliar to most readers, it should be clear that the flow rates and compositions of the process streams are not only important but also relatively simple to obtain.

工艺设计

工艺设计是利用管道将容器连接起来组成网络。计算每个容器和管道的流量显然是重要的，如此它们的大小才能被正确设定。这一设定不仅要针对正常的生产状况，而且也要针对开车阶段的状况以及一旦发生紧急停车阶段的状况。某些管道和容器专门针对后一目的来设计。虽然，人们希望没有必要去使用紧急系统，但对它的测试（特别是相关的控制设备）是必要的。

工厂中每部分的质量流速和体积流速的知识来源于物质平衡的计算。由于诸如这样的问题需要被回答，如“热量的作用是什么?”，“需要多少蒸汽?”,“我们有足够的冷却水吗?”,因而能量平衡也很重要。

对于一个由混合器、反应器和分离器组成的工艺工厂来说，平衡可逐个单元来计算。来自混合器的产物流是反应器里的一股填料流，同时来自反应器的产物流是分离器中的填料。然而，如果来自下游单元的流体返回到上游单元时，连续计算是不可能的。

无论什么时候需要循环未被转化的反应物时，就会出现循环问题。在哈伯法生产氨时，即使是接近平衡的转化也只能达到25%，因此氮气和氢气要被循环。反应物到产物的总转化率能达到接近100%的水平。用一氧化碳和氢气合成甲烷的过程中同样需要循环，因为转化率也在约25%的水平。用乙烯合成氯乙烯的生产也涉及循环流体。更多这样的例子可以列举出来。更多复杂的工艺常包括几个循环圈。在工艺设计阶段，不仅需要考虑循环圈在稳态操作状态下的效果，也要考虑在开车阶段的效果。

虽然大多数读者不熟悉以上所列举的工作，但工艺流体的流速和组成不仅重要而且相对容易获得，这一点应该是明晰的。

Unit 9

Reactor Type

The reactors are usually complex equipments. There are usually four broad kinds of reactors. They are stirred tank reactors, tubular reactors, packed bed reactors and fluidized bed reactors.

The stirred tank reactor is the basic chemical reactor. Tank sizes range from a few liters to several thousand liters. They are used for liquid-liquid reactions and liquid-gas reactions.

Tubular reactors are usually used for gaseous reactions. When small diameter tubes are used in the tubular reactors, it can increase heat transfer rate.

There are solids in the packed bed reactors. The solids may be a reactant or a catalyst. Packed bed reactors are used for gas and gas-liquid reactions.

The solids in a fluidized bed reactor are held in suspension by the reacting fluid. This promotes high mass transfer rates and heat transfer rates and good mixing.

New Words

complex [ˈkɔmpleks] *adj*. 复杂的，合成的，综合的
equipment [iˈkwipmənt] *n*. 装备，设备
tubular [ˈtjuːbjulə] *adj*. 管的，管形的
packed [ˈpækt] *adj*. 充满……的，塞满了……的
fluidized [ˈfluːidaizd] *adj*. 流态化的
liter [ˈliːtə] *n*. 公升
reactant [riːˈæktənt] *n*. 反应物
fluid [ˈflu(ː)id] *n*. 流体
promote [prəˈməut] *vt*. 促进，提升

Expressions and Technical Terms

stirred tank reactor 搅拌釜式反应器
tubular reactor 管式反应器
packed bed reactor 填料床式反应器
fluidized bed reactor 流化床式反应器
range from 在……范围内变化
liquid-liquid reaction 液-液反应
liquid-gas reaction 液-气反应
gaseous reaction 气体反应
reacting fluid 反应流体

mass transfer　质量传递

Exercises

A. Answer the following questions.

1. How many kinds of reactors usually are there? What are they?

2. What are stirred tank reactors used for?

3. What are tubular reactors usually used for?

4. What are packed bed reactors used for?

5. Can fluidized bed reactors promote mass transfer rates and heat transfer rates?

B. Translate the following into English.

1. 搅拌釜式反应器　　2. 管式反应器

3. 填料床反应器　　4. 流化床式反应器

5. 液-液反应　　6. 液-气反应

7. 气体反应　　8. 反应流体

9. 传质　　10. 传热

C. Reading comprehension. After reading a passage, choose the best answer to each question.

The atmosphere contains water vapors, but there is a limit in a given volume air, just as there is a limit to how much sugar can be dissolved in one cup of coffee.

More sugar can be dissolved in hot coffee than in cold. A given volume of air can hold more water vapor at a higher temperature than at a lower temperature. The air is said to be saturated when it holds as much water vapor as it can at that temperature. At 20℃ a cubic meter of air can hold about 17gm of water vapor; at 30℃ it can hold about 30gm. Usually the atmosphere is not saturated. If the temperature of the air drops sufficiently, saturation is reached and excess moisture precipitates out.

New Words and Phases

atmosphere [ˈætməsfiə] *n*. 大气，空气
vapor [ˈveipə] *n*. 水蒸气
volume [ˈvɔljuːm] *n*. 体积
excess [ikˈses] *adj*. 超过的，额外的
moisture [ˈmɔistʃə] *n*. 潮湿，湿气
cubic meter 立方米

1. When the air is at a higher temperature, it will contains ( )

   (A) less water vapor.　　(B) more water vapor.

2. If the temperature of the air drops sufficiently,

it is likely to be (　　)

(A) raining. (B) windy.

(C) getting warmer.

Reading Material

Process Reactor Design

工艺反应器设计

Process reactor design

- In the design of a process reactor, a chemical engineer must consider the following:
  1. What reaction will occur in the reactor?
  2. How fast could the reaction go?
  3. What type and size should the reactor be? What operating temperature, pressure, compositions, and flow rates should be selected?
  4. Is the production economical?
- The first question deals with the thermodynamics from which the equilibrium composition of the reaction mixture can be estimated. The second concerns the process kinetics from which the rate of

工艺反应器设计

- 在设计工艺反应器时，化学工程师必须考虑以下问题：
  1. 反应器里将发生的是什么反应？
  2. 反应的速率有多快？
  3. 反应器应该是哪种类型且有多大？应该选择什么样的操作温度、压力、组成和流速？
  4. 生产是否经济？
- 第一个问题与热动力学相关，从中可估计出反应混合物的平衡组成。第二个问题涉及过程动力学，从中可预测出反应的速率常数。第三个问题涉及反应系统中物质和能量平衡。第二和第三个

constant of reaction can be predicted. The third accounts for mass and energy balances in the reaction system. The incorporation of the second and third determines the type and size of the reactor required for certain reactions. The fourth question considers the economics of the process from which the optimum operating conditions can be obtained.

问题结合起来决定了某些反应所需的反应器种类和大小。第四个问题考虑的是工艺过程中的经济学方面，从中可得出最优的操作条件。

- In order to fulfill these requirements，we need information，knowledge，and experience from a variety of areas：thermodynamics，chemical kinetics，fluid mechanics，heat transfer，mass transfer，and economics.

- 为了达到这些要求，我们需要来自众多领域里的信息、知识和经验：如热动力学、化学动力学、流体力学、传热、传质和经济学。

Batch processing

间歇操作

- In a batch process all the reagents are added at the commencement. The reaction proceeds the compositions changing with time，and the reaction is stopped and the product withdrawn when the required conversion has been reached. Batch processes are suitable for small scale production and for

- 在间歇过程中，所有的反应物在开始阶段被加入。反应进行时，组成随时间变化。当达到所要求的转化时反应被停止，产物提取出来。间歇操作适合于小规模的生产，也适合于在同一设备中生产一范围内的不同产品，如颜料、染料和聚合物的生产。

processes where a range of different products is to be produced in the same equipment, for instance, pigments, dyestuffs and polymers.

Continuous processing

In a continuous processes, the reactants are fed to the reactor and the products withdrawn continuously. The reactor operates under steady state conditions. Continuous production will normally give lower production costs than batch production, but lacks the flexibility of batch production. Continuous reactors will usually be selected for large scale production. Processes that do not fit the definition of batch or continuous are often referred to as semi-continuous or semi-batch. In a semi-batch reactor some of the reactants may be added, or some of the products withdrawn, as the reaction proceeds. A semi-continuous process can be one which is interrupted periodically for some purpose. For instance, for the regeneration of catalysts.

连续操作

在连续操作中，反应物被连续填入到反应器中，产物被连续提取出来。反应器在稳定状态下操作。连续生产通常比间歇生产投入较低的生产成本，但缺少间歇生产的弹性。连续反应器通常被选择用于大规模生产。那些不符合间歇或连续生产定义的工艺过程通常被称为半连续或半间歇生产。在半间歇反应器中，随着反应的进行，一些反应物被加入，或一些产物被提取出。因某种原因，半连续过程能被定期地阻断。例如，催化剂的再生。

Unit 10

New Technologies in Unit Operation

New technologies are occurring all the time. The big one is the combination of chemical processes. When reaction, separation and heat exchange are combined, we can reduce the investment and reaction by-products.

With combined reaction/separation, we can also improve the yield for a reaction. When products are separated out, the feedstock cannot react with products. This also reduces waste of feedstock and product. Immediate removal of reaction products also will enable low equilibrium constant reactions to proceed much further. Indeed, reactions can be forced to 100% conversion through combined reaction/separation.

A second and newer technology is the minireactor. A plant may consist of many minireactors. The advantages include simpler monitoring and small-scale equipment.

New Words

combination [ˌkɔmbiˈneiʃən] *n.* 结合，联合，化合
investment [inˈvestmənt] *n.* 投资
feedstock [fiːdstɔk] *n.* 给料
immediate [iˈmiːdjət] *adj.* 直接的，紧接的，立即的
removal [riˈmuːvəl] *n.* 移动
equilibrium [ˌiːkwiˈlibriəm] *n.* 平衡
constant [ˈkɔnstənt] *n.* 常数
proceed [prəˈsiːd] *vi.* 进行，继续
minireactor [ˌminiriˈæktə] *n.* 小反应器
advantage [ədˈvɑːntidʒ] *n.* 优势，有利条件
include [inˈkluːd] *vt.* 包括，包含
monitor [ˈmɔnitə] *vt.* 监控

Expressions and Technical Terms

unit operation 单元操作
heat exchange 热量交换
combined reaction/separation 综合反应/分离
consist of 由……组成
equilibrium constant 平衡常数
small-scale equipment 小规模设备

Exercises

A. Fill in the blanks.

1. The big new technologies is ________ of chemical processes.

2. With combined reaction/separation, we can also improve ________ for a reaction.

3. When products are separated out, the ________ cannot react with ________.

4. ________ of reaction products also will enable low equilibrium constant reactions to proceed much further.

5. A second and newer technology is the ________.

B. Translate the following into English.

1. 单元操作　　2. 热交换　　3. 综合反应/分离
4. 平衡常数　　5. 小规模设备　　6. 小反应器
7. 原料

C. Reading comprehension. After reading a passage, choose the best answer to each question.

The atmosphere is a mixture of several gases. There are about ten chemical elements which remain in gaseous form under all natural conditions. Of these gases, oxygen makes up about 21 percent and nitrogen about 78 percent. Several other gases, such as argon, carbon dioxide, hydrogen, neon, krypton, and xenon, comprise the remaining one percent of the volume of dry air. The amount of water vapor is very important in weather changes.

The layer of the air next to the earth, which extends upward for about ten miles, is known as the troposphere. It makes up about 75 percent of all the weight of the atmosphere. It is the warmest part of the atmosphere because most of the solar radiation is absorbed by the earth's surface.

The upper layers are colder because of their greater distance from the earth's surface. It was assumed that upper air had little influence on weather changes. Recent studies have shown the assumption to be incorrect.

New Words

natural [ˈnætʃərəl] *adj*. 自然的，天生的

condition [kənˈdiʃən] *n*. 条件，环境

argon [ˈaːgɔn] *n*. 氩

neon [ˈniːən] *n*. 氖

krypton [ˈkriptɔn] *n.* 氪

xenon [ˈzenɔn] *n.* 氙

comprise [kəmˈpraiz] *v.* 包含，由……组成

extend [iksˈtend] *v.* 扩充，延伸

troposphere [ˈtrɔpəusfiə] *n.* 对流层

solar [ˈsəulə] *adj.* 太阳的，日光的

assume [əˈsjuːm] *vt.* 假定，设想

assumption [əˈsʌmpʃən] *n.* 假定，设想

1. The atmosphere consists of (　　)

(A) nine chemical elements.

(B) only nitrogen and oxygen.

(C) about 21 percent oxygen and about 78 percent nitrogen and other gases.

2. The troposphere is the warmest part of the atmosphere because (　　)

(A) most of the solar radiation is absorbed by the earth's surface.

(B) it is nearest the sun.

(C) it contains heat.

3. The upper layers are colder than the troposphere because of (　　)

(A) their most of the solar radiation.

(B) their greater distance from the earth's surface.

Reading Material

Chemicals from Crude Oil

Crude oil

Crude oil is a mixture. It consists of saturated hydrocarbons whose boiling points are close together. Crude oil or petroleum is a fossil fuel—the remains of marine organisms and is a non-renewable resource, but new reserves continue to be found.

Physical separation of mixture—fractional distillation

Fractions are groups of compounds with similar boiling points. By fractional distillation in the column, we can obtain different kinds of fractions such as petroleum gases, petrol, paraffin, diesel, light oil, fuel oil, wax and bitumen. These fractions can separately be used as bottled gas, jet fuel, truck, bus fuel, lubricants, heating fuel, candles ad roads.

Supply/demand mismatch problem

The demand for petrol and diesel is high and there is not a high enough proportion of these fractions in crude oil. The demand for fuel oil is low and there is more than enough of this fraction in crude oil. So the supply of fractions does not match the demand for them. Fuel oil is cracked to meet demand.

Chemical decomposition of compounds—cracking

The mismatch problem is solved by decomposing some of the fuel oil to make more petrol of diesel. This is called cracking. The large molecules are broken down or cracked either by heat (thermal cracking) or by catalyst and heat (catalytic cracking). Cracking also has the benefit of making reactive, unsaturated compounds with double bonds. These compounds are important petrochemical feedstocks. This means other useful substances can be made from them.

来自原油的化学品

原油

原油是一种混合物。它由沸点很接近的饱和烃组成。原油或石油是一种化石燃料——海洋有机物的残骸，也是一种不能再生的资源，但新的储存不断被发现。

用物理方法分离混合物——分馏

馏分是有相同沸点的一组化合物。通过塔中的分馏，我们能够得到不同种类的馏分如石油气、汽油、煤油、柴油、轻油、燃油、蜡和沥青。它们可用作瓶装天然气、汽车燃油、喷气飞机燃油、载重汽车、大客车燃油、润滑油、加热燃油、蜡烛和铺路。

供需矛盾

汽油和柴油需求量很大，但原油的馏分中，这两种馏分的比例不是很高。对燃油的需求却很低，而在原油里这种馏分的比例却很高。因此，馏分的组成无法满足实际需求。所以，要将燃油裂解以满足需求。

用化学方法分解化合物——裂解

通过分解一些燃油，以制取更多的汽油和柴油来解决供需矛盾。这种分解方法叫裂解。采用加热的方法（热裂解）或是采用催化加热的方法（催化裂解），大分子被打开或断裂。裂解还有一个好处，就是能制取较不活泼、不饱和的含双键的化合物。这些化合物是重要的石油化工原料，这意味着其他一些有用的物质可以从这些化合物中制取。

Unit 11

Crystallization, Precipitation and Filtration

If the water in a solution is lost by evaporation, the ions that are left will clump together. The positive ions will attract the negative ions so that oppositely charged ions are next to each other. A lattice will slowly form. The process is called crystallization.

Crystallization is used for the production, purification and recovery of solids. The process is widely used in chemical industries such as the production of pharmaceutical products, sugar, common salt and fertilizers.

Precipitation is the reaction between two solutions making an insoluble product. If two solutions containing oppositely charged ions which attract each other very strongly are added together, the ions will immediately clump together. Because this happens very quickly, there is not time for an ordered lattice to form. So a fine suspension is seen. This is called a precipitation.

For example:

$$CaCl_2 + Na_2CO_3 \longrightarrow CaCO_3 + 2NaCl$$

In filtration, a porous medium or screen is used. It can retain the solids from a suspension and allow the liquids to pass. In the chemical laboratory, filter paper is often used as a medium. Suspended particles are retained on the filter paper forming the residue. The smaller particles pass through the filter paper forming the filtrate.

New Words

crystallization [ˈkristəlaiˈzeiʃən] *n*. 结晶

precipitation [priˌsipiˈteiʃən] *n*. 沉淀

filtration [filˈtreiʃən] *n*. 过滤

evaporation [iˌvæpəˈreiʃən] *n*. 蒸发

clump [klʌmp] *v*. 使成团或成块，使结合

lattice [ˈlætis] *n*. 格子

recovery [riˈkʌvəri] *n*. 恢复，回收

pharmaceutical [ˌfɑːməˈsjuːtikəl] *adj*. 药物的

fertilizer [ˈfəːtiˌlaizə] *n*. 肥料

ordered [ˈɔːdəid] *adj*. 规则的，整齐的

porous [ˈpɔːrəs] *adj*. 多孔的

medium [ˈmiːdjəm] *n*. 媒介

retain [riˈtein] *vt*. 保持，保留

residue [ˈrezidjuː] *n*. 滤渣

filtrate [ˈfiltreit] *n*. 滤出液

Expressions and Technical Terms

a positive ion 正离子，阳离子

a negative ion 负离子，阴离子

common salt 食盐

filter paper 滤纸

Exercises

A. Answer the following questions.

1. How does a lattice form in crystallization?

2. What is crystallization used for?

3. What is the definition of precipitation?

4. What will happen if two solutions containing oppositely charged ions which attract each other very strongly are added together?

5. What is used to retain the solids from a suspension and allow the liquids to pass in filtration?

B. Translate the following into English.

1. 结晶　　2. 沉淀　　3. 过滤　　4. 多孔介质

5. 滤纸

C. Reading comprehension. After reading a passage, choose the best answer to each question.

Precipitation reactions can be used to test for both anions and cations. For example, the presence of chloride ions in solution can be detected by adding silver ions. Silver ions attract chloride ions so strongly that if there are any chloride ions present, a white precipitate of silver chloride will form. Equally, silver ions can be tested for using chloride ions.

Many metals form insoluble hydroxides. So if a solution of sodium hydroxide is added to a solution of the metal, a precipitate of the hydroxide will be seen. Hydroxide precipitates vary in color, and some redissolve when excess sodium hydroxide is added.

New Words

anion [ˈænaiən] *n*. 阴离子

cation [ˈkætaiən] *n*. 阳离子

presence [ˈprezns] *n*. 存在

detect [di'tekt] *vt*. 探测，测定
redissolve ['riːdi'zɔlv] *v*. 再溶解
silver chloride 氯化银

1. When you add some silver ions to the solution of chloride ions, what happens? ( )

(A) A white precipitate will be seen at the bottom of the solution.

(B) You can not see any precipitates.

2. If a solution of sodium hydroxide is added to a solution of the metal which can form insoluble hydroxide, ( )

(A) an acid-base reaction will take place.

(B) a precipitation reaction will take place.

Reading Material

Distillation

Distillation may be carried out by the two principle methods. The first method is based on the production of a vapor by boiling the liquid mixture to be separated and condensing the vapors without allowing any liquid to return to the still in contact with the vapors. The second method is based on the return of part of the condensate to the still under such conditions that this returning liquid is brought into intimate contact with the vapors on their way to the condenser. Both of the methods may be conducted as a continuous process or as a batch process.

Flash distillation is used most for separating components which boil at widely different temperatures. It is not effective in separating components of comparable volatility, since then both the condensed vapor and residual liquid are far from pure. By many successive redistillation small amounts of some nearly pure components may finally be obtained, but this method is too inefficient for industrial distillations.

A typical continuous fractionating column equipped with the necessary auxiliaries and containing rectifying and stripping sections. The plate on which the feed enters is called the feed plate. All plates above the feed plate constitute the rectifying section, and all plates below the feed, including the feed plate itself, constitute the stripping section. The feed flows down the stripping section to the bottom of the column in which a definite level of liquid is maintained. Liquid flows by gravity to reboiler. This is a steam-heated vaporizer which generates vapor and returns it to the bottom of the column. The vapor passes up the entire column. At one end of reboiler is a weir and the bottom product is withdrawn from the weir.

The vapors rising through the rectifying section are completely condensed in condenser and condensate is collect in accumulator. Reflux pump takes liquid from the accumulator and delivers it to the top plate of the tower. This liquid stream is called reflux. It provides the downflowing liquid in the rectifying section that is needed to act on the upflowing vapor. Without the reflux no rectification would occur in the rectifying section and the concentration of the overhead product would be no greater than that of the vapor rising from the feed plate. The overhead product is withdrawn form the product cooler.

蒸馏

蒸馏可通过两种基本的方法来完成。第一种方法是将待分离的液体混合物煮沸成蒸气，蒸气冷凝后没有任何液体返回到蒸馏器中与蒸气接触。第二种方法将部分冷凝物返回到蒸馏器中，这一回流液体与去往蒸馏器的蒸气发生密切接触。两种方法都可用作连续操作或间歇操作。

闪蒸主要用于沸腾时有很大温度差异的组分的分离。由于冷凝蒸气和残留液都远不能达到纯净，所以闪蒸对于有相近挥发度的组分的分离没有效果。虽然通过多次连续的二次蒸馏最终也可得到一些少量的几乎纯净的组分，但对于工业蒸馏而言，这种方法的效率太低。

典型的连续精馏塔装备有必要的辅助设备和精馏段和提馏段。料液从其上进入的板称为进料板。进料板上面所有的塔板组成了精馏段部分，进料板下面所有的塔板包括进料板构成提馏段部分。进料沿着提馏段流至塔底，塔底保持一定水平的液体。液体因重力流至再沸器，再沸器是一个蒸气加热蒸馏器，它使塔内液体产生蒸气并将该蒸气送至塔底。然后蒸气通过整个塔身。再沸器的一端是一个堰，塔底产品从其中取出。

上升并通过精馏段的蒸气在冷凝器中完全被冷凝，并收集在贮槽中。回流泵将贮槽内的液体输送到最顶部的塔板上，这一液体流叫作回流。它提供了精馏段内需要与上升蒸气相作用的向下流的液体。如果没有回流，精馏段就不会有精馏作用，塔顶产品的浓度也不会比自进料板上升的蒸气的浓度高。

Unit 12

The Production of Nitric Acid

In the production of nitric acid, the ammonia is converted into it. Nitric acid production consumes about 20% of all ammonia produced.

The conversion of ammonia to nitric acid is a three stage process.

(1) $4NH_3 + 5O_2 \longrightarrow 4NO + 6H_2O$

(2) $2NO + O_2 \longrightarrow 2NO_2$

(3) $3NO_2 + H_2O \longrightarrow 2HNO_3 + NO$

The first reaction is catalyzed by platinum and is carried out at about 900℃ in reactors. But at these temperatures, there are some important side reactions. By careful reactor design and by fine control of temperature, we can achieve a good effect.

In the second reaction, a mixture of air and nitrogen oxide is passed through a series of cooling condensers where partial oxidation occurs. The reaction is favored by low temperatures. In the third stage, the nitrogen dioxide is absorbed as it is passed down through a large absorption tower. Nitric acid emerges from the bottom.

New Words

convert [kən'vəːt] *vt*. 使转变，转换……
consume [kən'sjuːm] *vt*. 消耗
catalyze ['kætəlaiz] *vt*. 催化
platinum ['plætinəm] *n*. 白金，铂
achieve [ə'tʃiːv] *vt*. 完成，达到
condenser [kən'densə] *n*. 冷凝器
partial ['paːʃəl] *adj*. 部分的，局部的
oxidation [əksi'deiʃən] *n*. 氧化
favor ['feivə] *vt*. 促成，支持
emerge [i'məːdʒ] *vi*. 显现，形成

Expressions and Technical Terms

nitric acid 硝酸
convert into 使转变成……
side reaction 副反应
reactor design 反应器设计
nitrogen dioxide 二氧化氮
absorption tower 吸收塔

Exercises

A. Answer the following questions.

1. What catalyst is used in the production of nitric acid?

2. What is converted into nitric acid in the production?

3. how many stages are there in the production of nitric acid?

B. Translate the following into English.

1. 氨	2. 硝酸	3. 吸收塔
4. 副反应	5. 反应器设计	6. 冷凝器
7. 一氧化二氮	8. 铂	

C. Reading comprehension. After reading a passage, choose the best answer to each question.

Sulphuric acid is the chemical that is produced in the largest tonnage. The raw material for sulphuric acid production is elemental sulphur. Almost all the elemental sulphur produced is used in the manufacture of sulphuric acid.

The production of sulphuric acid has three stages.

(1) The burning of sulphur in air to give sulphur dioxide.

$$S+O_2 \longrightarrow SO_2$$

(2) The reaction of sulphur dioxide and oxygen to give sulphur trioxide.

$$2SO_2+O_2 \longrightarrow 2SO_3$$

(3) The absorption of sulphur trioxide in water to give sulphuric acid.

$$SO_3+H_2O \longrightarrow H_2SO_4$$

The first reaction is carried out at about 1000℃. Then the gas stream containing about 10% sulphur dioxide is cooled down to 420℃.

The second reaction is slow and requires a catalyst. In the third stage, the sulphur trioxide is hydrated with water to give acid solution in absorption tower.

A large amount of sulphuric acid is used in the production of phosphoric acid and ammonium sulphate for fertilizers.

New Words and Phases

tonnage [ˈtʌnidʒ] *n*. 吨位
hydrate [ˈhaidreit] *n*. 氢氧化物；*v*. 与水化合
fertilizer [ˈfəːtiˌlaizə] *n*. 肥料
phosphoric acid 磷酸
ammonium sulphate 硫酸铵

sulphur trioxide 三氧化硫

1. The raw material for sulphuric acid production is ( )

(A) sulphur trioxide.

(B) elemental sulphur.

(C) sulphur dioxide.

2. The acid solution is obtained in absorption tower in ( )

(A) the first reaction.

(B) the second stage.

(C) the third stage.

3. A large amount of sulphuric acid is used in the production of ( )

(A) phosphoric acid.

(B) ammonium sulphate.

(C) phosphoric acid and ammonium sulphate.

Reading Material

Hazards in Chemical Engineering Laboratories

For most students, working in a chemical engineering laboratory introduces them to operate on a much larger scale. Other changes are the increased quantity of material involved, the increased energy in systems, and the added complexity of dealing with runaway reactions.

Flammable liquids are used as reactants, solvents, or cleaning fluids. While safety may be improved by substitution, the strictest vigilance must be maintained on toxicity hazards. If flammable liquids are employed, the practical and physical significance of their properties should be explained to the users. The hazard should be minimized by limiting the quantity in the working area, providing adequate licensed storage and instituting a workable system of handling and withdrawal from the store. Sources of ignition must be eliminated. Coordination with other work in the vicinity is also important.

Students are unlikely to work with solid explosives but may become involved with oxidizing agents and combustible solids. These demand respect in handling and storage. The same applies to exothermic liquid phase reactions and "sealed-vessel" reactions. To avoid gas and vapor explosions it is necessary to avoid the confluence of flammable mixtures and an ignition source.

The student is likely to encounter several new aspects of the hazards of electricity; many are unfamiliar with the wiring of three-phase appliances and with the safe use of separate phases. All such wiring must be the responsibility of a trained electrician. High-powered immersion heaters require adequate cutout devices and fail-to-safety devices and earth leakage protection. Static electrical charges can present a hazard of ignition, shock or nuisance. Therefore adequate earthing or neutralization of charge should be incorporated into the apparatus.

Toxic materials are often used. If necessary, however, complete or partial containments should be used. Attention must then be directed to the quality and fate of any exhaust stream. Ventilation is a requirement under almost all circumstances. The hazard of sudden releases must be quantitatively assessed and large stocks of toxic material must be kept outside the laboratory.

It is concluded that hazards should be identified at the design stage by supervisor. Hazards can then be engineered-out. The effect of any which cannot must be quantified. This should be part of the normal training of any worker.

化学工程实验室中的危险

对大部分学生而言，在化工实验室工作将引导他们到一个更大规模的操作水平上。其他区别包括所涉及到的物质增加的量，系统中增加的能量，以及为处理失去控制的反应而增添的复杂性。

易燃液体可用作反应物、溶剂或清洗流体。虽然可以通过更换（易燃液体的品种）来提高安全性，但对于毒品的危险性仍须保持极高度的警惕。如果使用了易燃液体，那么它们性质上的实际意义和物理意义一定要说明给使用者。如果在工作场所限制（易燃物品的）储藏量，提供足够合格的储藏库，以及制订切实可行的处理和领取易燃物品的制度，那么危险性就可减少到最低程度。

学生不可能用到固体爆炸物但可能涉及到氧化剂和易燃固体，这些物质需要考虑它们的处理和储藏。同样的原理也适用于放热的液相反应和密闭容器内的反应。为了避免气体和蒸气爆炸，有必要防止易燃混合物和点火源的接触。

学生可能会遇到电危险的几个新方面；学生不熟悉三相装置的布线，也不熟悉每一相的安全使用。所有这些布线必须是受过训练的电工的职责。高动力的浸入式加热器需要足够的切断装置、安全保险装置和接地保护装置。静电荷将引起着火、电击或损害的危险，因此设备中应有适当的接地或中和电荷的设施。

有毒物质经常被使用，然而，如果必要的话，应对它们完全或部分限制使用。必须注意任何废气流的量和去向。几乎所有的环境都要求通风。毒物突然释放的危险必须进行定量的评价，大量有毒的物质必须保存在实验室外。

可以得出结论，危险应由监管人员在设计阶段就已鉴别出。然后，危险才能被巧妙地排除，对于任何无法排除的危险所引起的后果必须以量来表示。这应该是任何工作人员的部分常规训练。

Unit 13

Polymers

What are polymers? Polymers are complex and giant molecules. They are different from low molecular weight compounds like, say, common salt. The molecular weight of common salt is only 58. 5, while that of a polymer can be several hundred thousand. These big molecules are made up of many smaller molecules. For example, the molecular weight of butadiene, a gaseous compound, is 54. When it combines nearly 4000 times, we can get a polymer, polybutadiene, a synthetic rubber. Its molecular weight is about 200000.

How are polymers made? The polymer is made up of many small molecules. These small molecules are called monomer. When the monomer molecules are linked to form a big polymer molecule, the process is called polymerization. When molecules just add on to form the polymer, the process is called addition polymerization. However, when molecules do not just on but also undergo some reaction, the process is called condensation polymerization.

Classification of polymers polymers can have different chemical structures and physical properties. It can be classified in different ways.

(1) natural and synthetic polymers.

(2) organic and inorganic polymers.

(3) thermoplastic and thermosetting polymers.

(4) plastics, elastomers, fibers and liquid resins.

New Words

polymer [ˈpɔlimə] *n*. 聚合物

butadiene [ˌbjuːtəˈdaiiːn] *n*. 丁二烯

polybutadiene [ˌpɔliˌbjuːtəˈdaiiːn] *n*. 聚丁二烯

monomer [ˈmɔnəmə] *n*. 单体

polymerization [ˌpɔliməraiˈzeiʃən] *n*. 聚合

undergo [ˌʌndəˈgəu] *vt*. 经历，经过

condensation [kɔndenˈseiʃən] *n*. 浓缩

classification [ˌklæsifiˈkeiʃən] *n*. 分类，分级

classify [ˈklæsifai] *vt*. 分类，分等

synthetic [sinˈθetic] *adj*. 合成的

thermoplastic [ˌθəːməˈplæstik] *adj*. 热塑性的

thermosetting [ˌθəːməuˈsetiŋ] *adj*. 热固性的

plastic [ˈplæstik] *n*. 塑料

elastomer [iˈlæstəmə] *n*. 弹性体

fiber [ˈfaibə] *n*. 纤维

resin [ˈrezin] *n*. 树脂

Expressions and Technical Terms

be different from 与……不同

molecular weight 分子量

common salt 食盐

be made up of 由……组成

synthetic rubber 合成橡胶

be linked to 与……连接

add on to 增加，累加

addition polymerization 加聚反应

condensation polymerization 缩聚反应

chemical structure 化学结构

liquid resin 液体树脂

Exercises

A. Answer the following questions.

1. Are polymers a low molecular weight com-

pounds?

2. How much is the molecular weight of common salt?

3. How much is the molecular weight of polybutadiene about?

4. What is polymerization?

5. What is addition polymerization?

6. What is condensation polymerization?

B. Translate the following into English.

1. 聚合物　　2. 分子量　　3. 合成橡胶
4. 加聚聚合　　5. 缩聚聚合　　6. 单体
7. 天然聚合物　　8. 合成聚合物　　9. 有机聚合物
10. 无机聚合物　　11. 热固性聚合物
12. 热塑性聚合物　　13. 塑料　　14. 液体树脂

C. Reading comprehension. After reading a passage, choose the best answer to each question.

If carbon dioxide gas is generated in large enough quantities, it will collect in the atmosphere and cause an unwelcome warming effect on climate.

The decay of plants, including trees, generates carbon dioxide, but in forests such carbon dioxide is balanced by living plants, which give off oxygen. The use in industry of fuels derived from plants generates a large amount of carbon dioxide, such fuels include wood, coal and oil.

New Words and Phases

generate [ˈdʒenəˌreit] *vt*. 产生，发生
unwelcome [ʌnˈwelkəm] *adj*. 不受欢迎的
climate [ˈklaimit] *n*. 气候
decay [diˈkei] *vi*. 腐朽，腐烂
balance [ˈbæləns] *n*. 平衡
living [ˈliviŋ] *adj*. 活的
derive [diˈraiv] *vi*. 起源
conclude [kənˈkluːd] *vt*. 推断，断定
attributable [əˈtribjutəbl] *adj*. 可归于……的
contribute [kənˈtribjuːt] *v*. 贡献，促进
net [net] *adj*. 净余的
surplus [ˈsəːpləs] *adj*. 过剩的，剩余的
give off 发出（蒸汽、光等）

If the statements above are true, which of the following can properly be concluded from them?（　　）

（A）All of the carbon dioxide that can be generated in an industrial society is attributable to plants.

(B) An unwelcome warming effect on climate cannot be avoided, since carbon dioxide is given off by the natural processes of decay in plants.

(C) Forests contribute as much carbon dioxide to the atmosphere as does the industrial use of fuel derived from plants.

(D) A society that uses plant-based fuels in industry will contribute a net surplus of carbon dioxide to the atmosphere, unless the gas is reabsorbed in some way.

(E) No matter which fuels are used by industry, there is bound to be an increase in the total amount of carbon dioxide in the atmosphere and an unwelcome warming effect on climate.

Reading Material

Classification of Polymers 聚合物的分类

Plastics, elastomers, fibers and liquid resins 塑料、弹性体、纤维和液体树脂

- Depending on its ultimate form and use, a polymer can be classified as plastics, elastomers, fibers and liquid resins. When, for

- 依据聚合物的外形和用途，聚合物可划分为塑料、弹性体、纤维和液体树脂。例如，当利用热量和压力使聚合物被塑造成硬的坚

instance, a polymer is shaped into hard and tough articles by the application of heat and pressure, it is used as plastic. Typical examples are polystyrene, PVC.

韧的颗粒时，可用作塑料。典型的例子是聚苯乙烯、聚氯乙烯。

- When vulcanized into rubbery products exhibiting good strength and elongation, polymers are used as elastomers. Typical examples are synthetic rubber, silicone rubber.

- 当聚合物被硫化成橡胶产品展示出良好的强度和延长性时，聚合物被用作弹性体。典型的例子是合成橡胶、硅橡胶。

- If drawing into long filament-like materials, whose length is at least 100 times its diameter, polymers are said to have been converted into fibers.

- 如果被拉伸成细线状物质，长度至少是直径的 100 倍时，聚合物可被转变成纤维。

- Polymers used as adhesives, potting compounds, sealants, etc., in a liquid form are called as liquid resins.

- 聚合物被用作胶黏剂、灌封料、密封剂等液体形式时，被称为液体树脂。

Organic and inorganic polymers

有机和无机聚合物

- A polymer whose backbone chain is essentially made of carbon atoms is termed an organic polymers. The atoms attached to the backbone carbon atoms are usually those of hydrogen, oxygen, ni-

- 骨架链实质上是由碳原子组成的聚合物被称为有机聚合物。附着在骨架链碳原子上的通常是氢、氧和氮原子等。大部分合成聚合物是有机物，它们被广泛地研究着。事实上，有机聚合物的数量

trogen, etc.. The majority of synthetic polymers is organic and they are very extensively studied. In fact, the number and variety of organic polymers are so large that when we refer to polymers, we normally mean organic polymers.

和种类是如此丰富，以至于我们提到聚合物时，通常指的是有机聚合物。

- The molecules of inorganic polymers, on the other hand, generally contain no carbon atom in their chain backbone. Glass and silicone rubber are examples of inorganic polymers.

- 另一方面，无机聚合物的分子在它们的骨架链上通常不含碳原子。玻璃和硅橡胶是无机聚合物的例子。

Natural and synthetic polymers

天然和合成聚合物

- Depending on their origin, polymers can be grouped as natural or synthetic. Those isolated from natural materials are called natural polymers. For example, plants take in small molecules and build large molecules. Animals take in large molecules (by eating plants or other animals), break them down by digestion, then build new large molecules from them.

- 根据聚合物的来源，聚合物可分为天然的或合成的。那些从天然物中分离到的聚合物被称为天然聚合物。例如，植物吸收小分子，用小分子构成大分子。动物吸收大分子（食用植物或其他动物），通过消化，打碎这些大分子，然后形成新的大分子。

- Polymers synthesized from low

- 由低分子量化合物合成的聚合物

molecular weight compounds are called synthetic polymers. Typical examples are polyethylene, PVC (polyvinylchloride), nylon and terylene.

被称为合成聚合物。典型的例子是聚乙烯、聚氯乙烯、尼龙和涤纶。

Thermoplastic and thermosetting polymers

热塑性聚合物和热固性聚合物

- Some polymers soften when heated and harden when cooled. This process can be repeated as often as needed. These are called thermopl-astic polymers. PVC, polyethylene, nylon are examples of thermoplastic polymers.

- 有些聚合物当加热时会软化，冷却时会硬化，只要需要，这个过程可以反复进行。它们叫热塑性聚合物。聚氯乙烯、聚乙烯和尼龙是热塑性聚合物的例子。

- Other polymers soften when heated but then new bonds form between the chains and the polymer hardens permanently. It will not soften again. These are called thermosetting polymers.

- 有些聚合物加热时会软化，但是链之间形成了新键，聚合物就永久硬化，它不会再软化。它们叫热固性聚合物。

Unit 14

Surfactants

Surfactants consist of a hydrophilic and hydrophobic group. The hydrophobic group is normally a hydrocarbon. The hydrophilic group is polar groups.

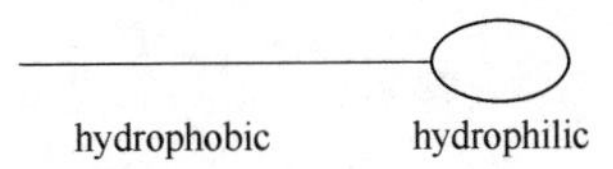

These compounds have a special property. For example, they can reduce the interfacial tension between water and the other phase. They can change wetting properties.

We can classify the surfactants into four categories.

Anionic surfactants,

Cationic surfactants,

Nonionic surfactants,

Amphoteric surfactants.

In anionic surfactant molecules, one hydrophobic hydrocarbon group is connected with one or two hydrophilic groups. The anion is the carrier of the surface active properties.

Cationic surfactants also contain a hydrophobic hydrocarbon group and one or several hydrophilic groups. However, the cation is the carrier of the surface active properties.

Nonionic surfactants do not dissociate into ions. Amphoteric surfactants can have anionic or cationic properties.

New Words

surfactant [səː'fæktənt] *n*. 表面活性剂

hydrophilic [ˌhaidrəu'filik] *adj*. 亲水的，吸水的

hydrophobic [ˌhaidrəu'fəubik] *adj*. 不易被水沾湿的，疏水的

hydrocarbon ['haidrəu'kɑːbən] *n*. 烃，碳氢化合物

phase [feiz] *n*. 相

category ['kætigəri] *n*. 种类

anionic [ˌænai'ɔnik] *adj*. 阴离子的

cationic ['kætaiənik] *adj*. 阳离子的

nonionic ['nɔnai'ɔnik] *adj*. 非离子的

amphoteric [ˌæmfə'terik] *adj*. 两性的

carrier [ˈkæriə] *n.* 运送者，携带者，载体

Expressions and Technical Terms

hydrophilic group 亲水基
hydrophobic group 疏水基
polar group 极性基
interfacial tension 界面张力
wetting property 润湿性
anionic surfactant 阴离子表面活性剂
cationic surfactant 阳离子表面活性剂
nonionic surfactant 非离子性表面活性剂
amphoteric surfactant 两性离子性表面活性剂
surface active property 表面活性性质
dissociate into 解离成……

Exercises

A. Answer the following questions.

1. What do surfactants consist of?

2. Can surfactants reduce the interfacial tension between water and the other phase?

3. Which part is the carrier of the surface active properties in anionic surfactants?

4. Which part is the carrier of the surface active properties in cationic surfactants?

5. Do nonionic surfactants dissociate into ions?

B. Translate the following into English.

1. 表面活性剂　　2. 亲水基　　3. 亲油基

4. 极性基团　　5. 界面张力

6. 阳离子型表面活性剂　　7. 阴离子型表面活性剂

8. 非离子型表面活性剂　　9. 两性离子型表面活性剂

C. Reading comprehension. After reading a passage, choose the best answer to each question.

You run into dozens of acids and bases every day. Orange juice contains acid. But the soap you use is a base. Believe it or not, there are acids and bases in your own body. Your stomach makes acid that helps you digest food. Acids are key ingredients in your muscles and skin. Your tears contain a base. So does your blood.

What makes an acid different from a base? The answer has to do with ions. When two chemical substances interact, atoms of one substance may lose negative charges to the other. The substance that loses a negative charge becomes a positive ion. The substance that gains a negative charge becomes a negative ion.

An acid is a substance that dissolves in water in a way that releases positive hydrogen ions (H^+) into the solution. A base is a substance that dissolves in water in a way that releases hydroxide ions (OH^-) in the solution. But ions are too small to see, even with a microscope. Are there other ways to tell if something is an acid or a base? Yes indeed!

Acids corrode metal. In many parts of the world, acid rain falls from the sky. When coal and other fossil fuels are burned, chemicals are released into the air. Inside the clouds, some of these chemicals combine with water to form acids. They fall to Earth as acid rain (or acid snow). Acids corrode metal structures. It also dissolves some kinds of stone and kills living things such as trees and fish. Acids taste sour.

Bases taste bitter. Bases feel slippery when you touch them. They are good at dissolving certain substances, like grease. That's one reason why bases make good cleaning agents. But bases also break down proteins. Some bases can actually dissolve your skin. Acids and bases are everywhere, but some of them need to be used with care.

New Words

juice [dʒuːs] *n*.（水果）汁，液

stomach [ˈstʌmək] *n*. 胃，胃部

digest [diˈdʒest] *v*. 消化

key [kiː] *n*. 关键

ingredient [inˈgriːdiənt] *n*. 成分

interact [ˌintərˈækt] *vi*. 互相作用，互相影响

corrode [kəˈrəud] *v*. 使腐蚀，侵蚀

fossil [ˈfɔsl] *adj*. 化石的

sour [ˈsauə] *adj*. 酸的，酸味的

bitter [ˈbitə] *adj*. 苦的

grease [griːs] *n*. 油脂

protein [ˈprəutiːn] *n*. 蛋白质

1. Your tears and your blood contains (　　)

(A) an acid.　　(B) a base.

2. The substance that loses a negative charge becomes (　　)

(A) a positive ion.　　(B) a negative ion.

(C) an atom.

3. When an acid dissolves in water, it can releases (　　)

(A) hydroxide ions.

(B) positive hydrogen ions.

4. Which of the following is true? (　　)

(A) Bases taste sour.

(B) Bases make good cleaning agents.

(C) Acid rain can not corrode metal.

Reading Material

Surfactants

- In solvents such as water, the surfactant molecules distribute in such a manner, that their concentration at the interfaces is higher than in the inner regions of the solution. They can orient themselves at the interface between the liquid and a solid, liquid, or a gaseous phase and modify the properties of the interface. This behavior is attributable to their amphiphilic structure (hydrophilic group, hydrophobic group).

表面活性剂

- 在水这样的溶剂中，表面活性剂以这种方式分布，即它们在界面处的浓度高于在溶液内部区域的浓度。它们能在液-固、液-液、液-气界面定向排列并改变界面性质。这一特点归因于它们的两性结构（亲水基、亲油基）。

- This definition makes surfactants sound academic, in fact the manufacture of soap, a major surfactant, is the oldest branch of chemical industry. Further more a survey of chemical industrial processes showed that surfactants were the most widely applied group of compounds in the chemical products industries. Surfactants are not only important as the active constituent of cleaning agents, which is their main use, but are also vital in the stabilization of emulsions (e. g. in foods and cosmetics), as mold release agents in the plastics industry, in flotation, in oil well drilling, and in a host of other applications.

- 这一定义使表面活性剂听起来较学术化，实际上，一种主要的表面活性剂——肥皂的生产是化学工业中最古老的一个分支。对化学工艺过程作进一步的了解，表明在化学产品工业中表面活性剂是一组最广泛使用的化合物。表面活性剂不只是作为洗涤剂的活性组分——这当然也是其主要应用，同时在如乳化液的稳定（如食品、化妆品中）、塑料工业中脱模、采矿工业中浮选、采油钻井及许多方面都有应用。

- Currently the most important synthetic anionic surfactant is sodium linear dodecylbenzene sulfonate (LAS). It is made by sulfonation of linear dodecylbenzene, which is made by alkylation of benzene with a straight chain dodecene whose double bond is

- 目前最重要的合成阴离子表面活性剂是直链的烷基苯磺酸钠。它是由线性十二烷基苯经磺化制成。十二烷基苯则由苯与直链十二烯经烷基化制得，十二烯并不要求双键在链端。

not necessarily in the terminal position.

All industrially important cationics are fatty nitrogen compounds and many are quaternary nitrogen compounds.

实际上工业中所有的重要的阳离子表面活性剂都是含氮化合物，其中许多是季铵盐类化合物。

Acronymofkey surfactants

重要的表面活性剂缩写

Acronym 缩写	Chinese name 中文名	Chemical name 化学名	类　型
LAS	直链烷基苯磺酸盐	Linear alkylbemzene sulphonate	Anionic surfactant 阴离子型表面活性剂
AS	烷基磺酸盐	alkane sulphonate	
AOS	α-烯烃磺酸盐	α-olefin sulphonate	
FAS	脂肪醇磺酸盐	fatty alcohol sulfate	
QAC	季铵化合物	quaternary ammonium compound	Cationic surfactant 阳离子型表面活性剂
AEO	脂肪醇聚氧乙烯醚	fatty alcohol polyoxyethylene ether	Nonionic surfactant 非离子型表面活性剂
APEO	烷基酚聚氧乙烯醚	alkylphenol polyethyleneglycol ether	
CAPB	椰油酰胺丙基甜菜碱	cocamidopropyl betaine	Amphoteric surfactant 两性离子型表面活性剂

Unit 15

Detergents

The detergent is a complex chemical product. There are many different kinds of substances in it. These ingredients may be classified into four major categories:

Surfactants;

Builders;

Bleaching agents;

Auxiliary agents.

1. Surfactants

Surfactants are the most important group. They have many functions.

Adsorption;

Soil removal;

Low sensitivity to water hardness;

Wetting power;

Good solubility.

2. Builders

Builders have a number of functions. They remove metal ions like calcium and magnesium. They can enhance the cleaning effectiveness.

3. Bleaching agents

There are two kinds of bleaching agents, the peroxide and the hypochlorite. They can achieve a satisfactory cleaning effect.

4. Auxiliary agents

(1) Enzymes.

(2) Fluorescent whitening agents. (FWA)

(3) Foaming stabilizers.

New Words

detergent [di'təːdʒənt] *n*. 清洁剂，去垢剂

builder ['bildə] *n*. 增洁剂

auxiliary [ɔːg'ziljəri] *adj*. 辅助的

function ['fʌŋkʃən] *n*. 官能，功能

adsorption [æd'sɔːpʃən] *n*. 吸附

soil [sɔli] *n*. 污垢

sensitivity ['sensi'tiviti] *n*. 敏感，灵敏性

remove [ri'muːv] *v*. 移动，去除

enhance [inˈhɑːns] *v*. 提高
effectiveness [iˈfektivnis] *n*. 效力，有效性
peroxide [pəˈrɔksaid] *n*. 过氧化物
hypochlorite [ˌhaipəuˈklɔːrait] *n*. 次氯酸盐
enzyme [ˈenzaim] *n*. 酶
fluorescent [fluəˈresənt] *adj*. 荧光的，发荧光的

Expressions and Technical Terms

be classified into 分类为……
bleaching agent 漂白剂
auxiliary agent 辅助剂
water hardness 水硬度
wetting power 润湿力
metal ion 金属离子
fluorescent whitening agent（FWA） 增白剂
foaming stabilizer 起泡稳定剂

Exercises

A. Answer the following questions.

1. What kinds of substances are there in detergents?

2. What are the functions of surfactants?

3. What are the functions of builders?

B. Translate the following into English.

1. 洗涤剂	2. 漂白剂	3. 润湿力
4. 水硬度	5. 助洗剂	6. 辅助剂
7. 荧光增白剂	8. 稳定剂	9. 酶
10. 过氧化物	11. 次氯酸盐	

C. Reading comprehension. After reading a passage, choose the best answer to each question.

Acids and bases can be dangerous, how can you figure out if the liquids are acids or bases without touching, or smelling them? The easiest way is to use an indicator. An indicator is something that changes color when it is exposed to an acid or a base. One of the most common acid/ base indicators is litmus paper. This is paper that has been soaked in a mixture of chemicals called litmus. Litmus may be the world's oldest acid/base indicator. It has been used for hundreds of years. Litmus paper has two forms—blue and red. If you dip a strip of blue litmus paper into a solution and it turns red, you know you have got an acid. If you dip a strip of red litmus paper into a solution and it turns blue, you have got a base. It is that simple.

But there is something more you need know. How strong are the acids or the bases? If you spilled apple juice on your skin, you would not worry. But if you spilled sulfuric acid on your skin, you would get a terrible burn. Both these liquids are acids. But there is a big difference between them. Sulfuric acid is much more powerful, or stronger acid than apple juice.

Whether an acid is strong or weak depends on how many hydrogen ions it releases when it dissolves in water. The more hydrogen ions there are in the solution, the stronger the acid is. Similarly, the strength of a base depends on how many hydroxide ions are in the solution. The more hydroxide ions there are, the stronger the base is. Scientists measure the strength of an acid or a base. This measurement is called the pH. A set of pH number is from 0 to 14. Pure water is in the middle, with a pH of 7. That means pure water is neutral, which is neither an acid nor a base.

What happens when you mix equal amounts of an acid and a base of equal strength? They cancel each other to make a neutral solution. Scientist calls this reaction "neutralizing."

New Words and Phases

indicator ['indikeitə] *n.* 指示剂

expose [iks'pəuz] *v.* 使暴露，揭露

litmus ['litməs] *n.* 石蕊

soak [səuk] *v.* 浸，泡，浸透

strip [strip] *n.* 条，带

spill [spil] *v.* 溢出，溅出

neutral ['nju:trəl] *adj.* 中性的

figure out 计算出

litmus paper 石蕊试纸

1. When a strip of blue litmus paper turns red in a solution, that means the solution is (　　)

(A) an acid solution.　　(B) a base solution.

(C) a neutral solution.

2. Sulfuric acid is (　　)

(A) much more powerful than apple juice.

(B) weaker than apple juice.

3. When there are more hydrogen ions in the solution, it means (　　)

(A) the weaker the acid is.

(B) the stronger the acid is.

(C) the solution is a neutral solution.

Reading Material

Detergent Ingredients and Its Formulations

洗涤剂的成分及其配方

Builders

The essential builder substances are follows.

- **alkalis (e. g. sodium carbonate, sodium silicate).**

  Sodium carbonate and sodium silicate were the only detergent components at the beginning of the 20th century. They precipitated the metal ions, which contributed to water hardness. They increased the pH value of the detergent solution. At the same time, they caused detrimental deposits of alkaline earth carbonates and silicates on fabrics and washing machine elements.

- **sequestrants or complexing agents (e. g. sodium triphosphate, STP).**

  Sodium triphosphate holds a dom-

助洗剂

基本的助洗剂物质如下。

- **碱(如碳酸钠、硅酸钠)**

  20世纪初，碳酸钠和硅酸钠是洗涤剂中仅有的成分。它们沉淀出导致水硬度的金属离子。它们增加了洗涤液的pH值。同时，它们也导致碳酸碱金属和硅酸碱金属在织物和洗衣机元件上有害的沉积。

- **螯合剂和络合剂(如三磷酸钠，STP)**

  在配位剂类的洗涤剂中，三磷酸

inant position among the complexing agents for detergents. This compound has a number of advantages over alkalis. A significant advantage is that the ions, which contribute to water hardness, are not precipitated but are eliminated through sequestration in the form of soluble complexes.

钠占主要的地位。这种化合物比碱具有许多的优点。一个显著的优点是导致水硬度的金属离子不是被沉淀出来而是通过螯合作用以可溶的配合物形式被去除。

- **ion exchangers.**

In order to control eutrophication, legal restriction on the phosphate content in detergents have been imposed in many states. Zeolite A is a suitable phosphate substitute. Zeolite A is manufactured from raw materials easily accessible throughout the world. It has turned out to be absolutely harmless with respect to environmental and toxicological effects.

- **离子交换剂**

为了控制富营养化作用，限制磷酸盐在洗涤剂中的含量这一法律规定在许多国家已被执行。沸石A是磷酸盐合适的替代品。沸石A可从世界各地容易得到的原材料中生产出。针对环境和毒理学效应来说，沸石A被证明完全无害。

Bleaching systems

Bleaching in general is any color change of a body towards decolorization, i. e. , an increase of the emission of visible light. Strictly speak-

漂白系统

漂白通常指的是通过脱色使物体任何颜色的变化，即可见光发光的增加。严格来说，化学漂白指的是颜色系统的氧化降解或还原降

ing, chemical bleaching designates the oxidative or reductive degradation of colored systems. With regard to the oxidizing bleach process two types of bleach, the peroxide bleach and the hypochlorite bleach, have become dominant in household washing.

解。针对氧化漂白过程，过氧化物和次氯酸盐漂白这两种漂白已经在家用洗涤中占主要地位。

Because of the specific laundering habits and in particular because of the higher laundry temperatures which are preferred in Europe, peroxide bleach predominates in Europe.

因为特殊的洗涤习惯，特别因为在欧洲更被喜欢用较高的洗涤温度，所以过氧化物漂白在欧洲占主要地位。

Lipstick formulations

	(1) Percent /%	(2) Percent /%
Caster oil	30.0	50.0
Mineral oil	10.0	—
Beeswax	12.3	7.0
Paraffin	10.0	—
Carnauba wax	10.0	3.0
Ceresin wax	10.0	3.0
Silicone fluid	10.0	—
Lanolin	—	10.0
Isopropyl myristate	—	5.0
Candelilla wax	—	7.0

唇膏配方

	(1) 百分数 /%	(2) 百分数 /%
蓖麻子油	30.0	50.0
矿物油	10.0	—
蜂蜡	12.3	7.0
石蜡	10.0	—
加诺巴蜡	10.0	3.0
地蜡	10.0	3.0
硅流体	10.0	—
羊毛脂	—	10.0
肉豆蔻酸异丙酯	—	5.0
小烛树蜡	—	7.0

p-hydroxybenzoic acid propyl ester	0.2	0.2	对羟基苯酸丙基酯	0.2	0.2
Bromoacetic acids	1.5	3.0	溴乙酸	1.5	3.0
Perfume	6.0	11.8	香料	6.0	11.8

Typical shampoo formulations

Ingredients	Wt./%	Function
Clear liquid		
Sodium lauryl sulfate	40.0	Cleansing agent
Lauramide	4.0	Foam stabilizer
Disodium EDTA	0.1	Sequestering agent
Formaldehyde	0.04	Preservative
Fragrance	0.5	Fragrance
FD&C blue No. 1	0.001	Color
FD&C yellow No. 1	0.004	Color
Deionized or distilled water	55.355	
Pearlescent		
TEA lauryl sulfate	20.0	Cleanser
Sodium lauryl sulfate	20.0	Cleanser
Cocoamide	5.0	Foam stabilizer
Glycol stearate	1.0	Opacifier, pearlescent agent
Disodium EDTA	0.1	Sequestering agent
Methylparaben	0.1	Preservative
Propylparaben	0.01	Preservative
Fragrance	0.5	Fragrance
Deionized or distilled water	53.29	

典型的香波配方

成　　分	质量分数/%	功　　能
澄清液		
十二烷基硫酸钠	40.0	清洗剂
十二酰胺	4.0	泡沫稳定剂
依地酸二钠	0.1	螯合剂
甲醛	0.04	防腐
香料	0.5	香料
蓝色一号铝沉淀颜料	0.001	颜色
黄色一号铝沉淀颜料	0.004	颜色
去离子水或蒸馏水	55.355	
珠光剂		
十二烷基硫酸三乙醇胺	20.0	净洗剂
十二烷基硫酸钠	20.0	净洗剂
可可酰胺	5.0	泡沫稳定剂
乙二醇单硬脂酸酯	1.0	遮光剂,珠光剂
依地酸二钠	0.1	螯合剂
尼泊金甲酯	0.1	防腐
尼泊金丙酯	0.01	防腐
香料	0.5	香料
去离子水或蒸馏水	53.29	

Unit 16

Coatings

Coatings are used to give long-term protection. They can used in chemical plants, paper plants, food plants, cars, bicycles, refrigerators and ships, etc.

Coatings can have a resistance to continuing exposure to chemicals, water or seawater, weather and high humidity. In order to achieve these functions, coatings must be inert and dense, must have a resistance to the transfer of chemicals through the coating, must be able to expand and contract, must be impervious to air, oxygen, water, carbon dioxide, ions, etc.

New Words

coating [ˈkəutiŋ]　*n*. 涂料

refrigerator [riˈfridʒəreitə]　*n*. 电冰箱

resistance [riˈzistəns]　*n*. 抵抗，抵抗力

exposure [iksˈpəuʒə]　*n*. 暴露，揭露

humidity [hjuːˈmiditi] *n*. 湿气，潮湿
dense [dens] *adj*. 密集的，浓厚的
expand [iksˈpænd] *vt*. 使膨胀，扩张
contract [ˈkɔntrækt] *v*. 收缩
impervious [imˈpəːvjəs] *adj*. 不可渗透的，透不过的

Expressions and Technical Terms

long-term protection 长期保护
paper plant 造纸厂
food plant 食品厂
be impervious to 对……不可渗透的
carbon dioxide 二氧化碳

Exercises

A. Fill in the blanks.

1. Coatings are used to give ________ protection.

2. Coatings can have a ________ to continuing exposure to chemicals, water or seawater.

3. Coatings must be ________ and dense, must have a ________ to the transfer of chemicals through the coating.

4. Coatings must be ________ to air, oxygen, water, carbon dioxide, ions, etc.

B. Translate the following into English.

1. 涂料　　2. 长期保护　　3. 纸厂
4. 食品厂　　5. 二氧化碳

C. Reading comprehension. After reading a passage, choose the best answer to each question.

Different kinds of matter have different physical and chemical properties. The properties of a substance are its characteristics. We know one substance from another by their physical and chemical properties. In a physical change, the composition of a substance is not changed. For example, ice can be changed into water. This is a physical change because the composition of a substance is not changed. In a chemical change the composition of a substance is changed. One or more new substances are formed.

Iron rusts in moist air. When iron rusts, it react with the oxygen from the air. A new substance is formed. It is called iron oxide. It has other different properties. Wood will burn if it heated in air. When wood burns, it reacts with the oxygen from the air. New substances are formed. They are carbon dioxide and water. Carbon dioxide and water have different properties. Heat is given off if the combustion of any fuel takes place. The above two cases are chemical changes.

New Words

composition [kɔmpəˈziʃən] *n*. 成分，组成
rust [rʌst] *n*. 铁锈 *vt*.（使）生锈
moist [mɔist] *adj*. 潮湿的 *n*. 潮湿
case [keis] *n*. 案例，情形

1. When ice is changed into water, it occurs (　　)

(A) a chemical change.

(B) a physical change.

(C) both a chemical and physical change.

2. When iron rusts in moist air, we can say that (　　)

(A) a physical change occurs.

(B) iron react with the oxygen from the air.

(C) no other new substances are formed.

Reading Material

A Resistant Coating and Lining

抗性涂料和涂层

A resistant coating

A resistant coating is a film of material applied to the exterior of structure steel, tank surface, conveyor band, pipes, process equipment or other surfaces which is subject to condensation, fumes, dusts or splash, but is not necessarily subject to immersion in any liquid or chemical. The coating must prevent corrosion of the structure by the environment.

抗性涂料

抗性涂料是一层用在结构钢外部、釜表面、传送机带、管道、工艺设备或其他表面的物质膜。它会遇到压缩、烟雾、灰尘或喷淋的环境，但不会浸泡到任何液体或化学品中。涂料必须能防止设备因环境而导致的腐蚀。

A resistant lining

A resistant lining is a film of material applied to the interior of pipe, tanks, containers or process equipment and is subject to direct contact and immersion in liquids, chemicals, or food products. As such, it must also prevent corrosion

抗性涂层

抗性涂层是一层用在管道、釜、容器或工艺设备外部的物质膜，它与液体、化学品或食品的直接接触并浸泡在其中。如此，涂层必须要阻止所装载的产品对设备的腐蚀，也要防止对产品的污染。就涂层而言，防止产品的污染是它最

of the equipment by the contained product, but must also prevent contamination of it. In the case of a lining, preventing product contamination may be its most important function.

重要的功能。

The function of a resistant coating or lining

抗性涂料或涂层的功能

The function of a high performance coating or lining is to separate two highly reactive materials, i. e. , to prevent strongly corrosive industrial fumes or liquids, solids, or gases from contact with the reactive equipment or underlying surface. The concept that a coating is a very thin film separating two highly reactive materials brings out the vital importance of the coating and its need to be completely continuous in order to fulfill its function. Any imperfection in the coating becomes a focal point for corrosion and breakdown of the equipment or focal point for the contamination of the contained liquid. The relatively thin, continuous film concept takes on

高性能的涂料或涂层就是将两种高反应性的物质分隔开，即阻止强腐蚀性的工业烟雾或液体、固体、气体与可反应的设备或内表面接触。涂料是一层分隔开两种高反应性物质的很薄的膜，这一概念显示出涂料的重要性。为了履行这一功能，涂料必须是完全连续的。涂料的任何缺陷都将成为设备腐蚀和损坏的关键点，或是所装载的液体受到污染的关键点。当这些保护性的涂料被用到很大面积的结构钢、釜表面和相似区域时，这一相对薄的连续膜的概念呈现出更大的重要性。

even greater importance when these protective coatings are applied to very large areas of structural steel, tank surfaces and similar areas.

Inhibitive coating

Inhibitive coating perform best in areas where the coating is subject to high humidities or chemical fumes. Such uses are generally for the exterior maintenance of equipment where the coating may be subject to physical abrasion and coating damage.

防锈涂料

防锈涂料在高湿度或化学烟雾的区域使用时最好。这些用途通常针对设备的外部维护，在这些区域，涂料可能会受到物理磨损和破坏。

Unit 17

Air Pollutants

When air contains one or more chemicals and harms humans, other animals, plants or materials, the phenomenon is called air pollution.

There are two major types of air pollutants.

A primary air pollutant is a chemical added directly to the air and it occurs in a harmful concentration. For example, when carbon dioxide is above its normal concentration, it is considered to be a air pollutant. A lead compound emitted by cars is also the air pollutant.

A secondary air pollutant is formed in the atmosphere through a chemical reaction.

Major air pollutions are as following:

1. Carbon oxides: carbon monoxide (CO), carbon dioxide (CO_2).

2. Sulfur oxides: sulfur dioxide (SO_2), sulfur trioxide (SO_3).

3. Nitrogen oxides: nitrogen dioxide (NO_2), dinitrogen oxide (N_2O), nitrogen oxide (NO).

4. Hydrocarbons: methane (CH_4), butane (C_4H_{10}), benzene (C_6H_6).

5. Smoke, dust.

6. Inorganic compounds: hydrogen fluoride (HF), hydrogen sulfide (H_2S), ammonia (NH_3), sulfuric acid (H_2SO_4), nitric acid (HNO_3).

7. Noise.

New Words

pollutant [pəˈluːtənt] *n*. 污染物
harm [hɑːm] *vt*. 伤害，损害
phenomenon [fiˈnɔminən] *n*. 现象
primary [ˈpraiməri] *adj*. 主要的，初步的，初级的
concentration [ˌkɔnsenˈtreiʃən] *n*. 浓度
emit [iˈmit] *vt*. 发出，散发
secondary [ˈsekəndəri] *adj*. 二级的，中级
methane [ˈmeθein] *n*. 甲烷，沼气
butane [ˈbjuːtein] *n*. 丁烷
benzene [ˈbenziːn] *n*. 苯
dust [dʌst] *n*. 灰尘，尘土

Expressions and Technical Terms

air pollution 空气污染
carbon monoxide 一氧化碳
sulfur trioxide 三氧化硫
nitrogen dioxide 二氧化氮
dinitrogen oxide 一氧化二氮
inorganic compound 无机化合物
hydrogen fluoride 氟化氢
hydrogen sulfide 硫化氢

Exercises

A. Answer the following questions.

1. What does air pollution mean?

2. What is the definition of primary air pollutant?

3. What is the definition of secondary air pollutant?

B. Translate the following into English.

1. 一次大气污染物 2. 二次大气污染物

3. 空气污染 4. CO_2

5. SO_3 6. NO_2

7. N_2O 8. CH_4

9. 苯　　　　10. H_2SO_4

C. Reading comprehension. After reading a passage, choose the best answer to each question.

Many reactions are reversible. In order to carry out a reversible reaction efficiently, the important conditions are that the reactants should be converted into products to the upmost extent and that this conversion should be brought about in the shortest possible time. For example, consider the reaction:

$$N_2 + 3H_2 \rightleftharpoons 2NH_3 + heat$$

In order to obtain a large yield of ammonia from a given quantity of nitrogen and hydrogen, the left-to-right reaction must be as nearly complete as possible. It has been found by experiment that a large yield of ammonia can be obtained under a high pressure, at a reasonable temperature and in the presence of a catalyst.

Heat is given off in the reaction, so the reaction will become more nearly complete at comparatively low temperatures. Yet the trouble is that too low a temperature will lower the reaction rate. Therefore, we must select a reasonable temperature. In addition to it, we must use a catalyst to speed up the left-to-right reaction.

New Words and Phases

reversible [ri'vəːsəbl] *adj*. 可逆的

upmost ['ʌpməust] *adj*. 最高的，最上的

reasonable ['riːznəbl] *adj*. 合理的，有道理的

carry out 实现，完成，执行

speed up 加速

1. In the production of ammonia, we must use a catalyst because ( )

(A) it can speed up the left-to-right reaction.

(B) it can lower the reaction temperature.

2. In order to obtain a large yield of ammonia, we must select ( )

(A) a very high temperature.

(B) a reasonable temperature.

(C) a very low temperature.

Reading Material

Environmental Issues

For the first two thirds of the twentieth century, the environmental effects of energy supply and use have been of minor concern. However, in the twenty-first century, the impact of energy generation and use on the environment has become a major concern. It is a major factor influencing the development of new technologies. How energy use will affect the environment is also an essential concern in making plans to meet future energy needs.

In developing countries, new industries require increasing amounts of energy production and use. Often coal is the least expensive and most available fuel for development. Alternative fuels are more expensive and technologies to control emissions add to the cost of energy as well. World population is projected to reach an estimated 10 to 15 billion people in the mid-twenty-first century. New technologies will be needed to minimize polluting wastes and to use energy efficiently in an even more crowed world. At the same time, energy conservation is becoming increasingly important. Predicting future energy needs is essential to meeting them. How energy will be supplied, stored, converted, and used must be considered.

Power plants will increasingly need to rely on renewable energy sources as nonrenewable resources run out. Global electric generation is projected to triple over the next thirty years.

The majority of today's environmental problems on a local, regional, or global scale are linked to worldwide energy supply and use. These problems are made worse because of the strong reliance on fossil fuels as primary energy sources. As the costs of depleting fuel resources are made clear, new energy sources will need to be developed to replace fossil fuel use. If fossil fuels continue to be used at the present rate, the proven reserves of gas and oil will be used up by around 2030, with other possible resources that might extend that time to a hundred years or more. Known coal reserves should last about four hundred years, with other possible coal resources likely to last several thousand years. Each year, though, an amount of fossil fuel is used that required about a million years to produce. Eventually, the supply must run out. It is possible that emissions due to fossil fuel may be found to affect human health or the world's ecological balance more than is now suspected. In that case, energy conservation and efficiency—as well as alterative fuels—will become even more important.

环境问题

在21世纪前三分之二的时期内，人们很少关注能量供给与使用对环境的影响。然而，进入21世纪以来，这一问题已成为人们关注的焦点。它也是影响新技术开发的主要因素之一。同时，人们在制定计划满足未来能源需求时，还必须着重考虑能源利用将对环境产生的影响。

在发展中国家，新兴产业对能源产生与使用的需求日益增大。通常来讲，煤的开采成本最低，也是最易获得的燃料。其他燃料成本则相对较高，又需某些技术以控制其废物排放，从而使用能量的成本大大增加。预计到21世纪中叶，世界人口将达到约100亿至150亿。世界将变得更为拥挤，因此人们需要一些新技术，以最大限度地减少低污染性废物。同时，能源保护也变得日益重要。要满足未来能源需求，相关的预测至关重要。人们必须考虑将如何供应、储存、转化与使用能源。

随着不可再生资源的枯竭，发电厂将日益需要依赖可再生能源，在未来30年内，全球发电总量计划增长两倍。

现今大部分环境问题，不管是地方性的、地区性的，还是全球性的，都与世界范围内能源的供应与使用有关。而人们对矿物燃料严重依赖，将其作为主要能源，使这些问题变的愈加严峻。与此同时，矿物燃料资源正趋向衰竭。使用这些资源的代价越来越清晰，人们需要开发新型能源以取而代之。按矿物燃料现在的使用速率计算，现已探明的油气储量将在2030年左右耗尽；如发现其他可能的储量，这一时间可能延长至100年或更长。问题是每年消耗的大量矿物能源需要100万年的漫长等待才能形成。因此，其储量将最终耗尽。另外，人们已开始怀疑矿物燃料利用过程中排放的废物会对人类健康与地球生态平衡有所影响，而这种影响最终可能要比现在人们所想的更为严重。这样一来，加强能源保护、提高其利用效率并寻找其他替代燃料将变得更加重要。

Unit 18

Titration

Titration is one of the most useful and accurate analytical techniques. It is fairly rapid. In a titration, the test substance (analyte) reacts with a solution of known concentration. This is called a standard solution, and it is generally added from a buret. The added solution is called titrant. The volume of titrant, which just completely reacts with analyte is measured. Since the concentration is known and since the reaction is known and since the reaction is known, the amount of analyte can be calculated.

There are many requirements in a titration.

(1) There is a known reaction between the analyte and the titrant. For example,

$$CH_3COOH + NaOH \longrightarrow CH_3COONa + H_2O$$

(2) The reaction should be rapid. Most ionic reactions are very rapid.

(3) There are no side reactions. If there are inter-

fering substances, these must be removed. In the above example, there should be no other acids.

(4) When the reaction is complete, there is a marked change in some property of the solution.

New Words

titration [tai'treiʃən] *n.* 滴定

accurate ['ækjurit] *adj.* 正确的，精确的

analyte ['ænəlait] *n.*（被）分析物

buret [bjuə'ret] *n.* 滴定管，玻璃量管

titrant ['taitrənt] *n.* 滴定剂

calculate ['kælkjuleit] *v.* 计算

Expressions and Technical Terms

analytical technique 分析技术

known concentration 已知浓度

standard solution 标准溶液

ionic reaction 离子反应

side reaction 副反应

interfering substance 干扰物质

Exercises

A. Fill in the blanks.

1. Titration is one of the most useful and accurate ______.

2. Standard solution is generally added from a ________.

3. In a titration, there should be a ________ between the analyte and the titrant.

4. In a titration, the reaction should be ________.

5. There are no side reactions. If there are interfering substances, these must be ________.

B. Translate the following into English.

1. 滴定　2. 滴定剂　3. 副反应
4. 离子反应　5. 标准溶液　6. 已知浓度
7. 干扰物质　8. 分析物

C. Reading comprehension. After reading a passage, choose the best answer to each question.

Fluids are often caused to move through pipes by pumps. The chief use of pumps is to add energy to the fluid. Energy can be used to do work. The energy to be added will serve to raise the pressure, elevation and velocity.

The terms "air pumps" and "vacuum pumps" are used to name the machines to compress a gas, but commonly pumps are regarded as devices to handle liquids.

It is not possible for a pump to work by itself. Therefore an engine is employed to make the pump run. Pumps are of great importance in industry. Various kinds of pumping equipment are widely used in industry. There are too many pumps. However, they can generally be divided into four classes: centrifugal pumps, reciprocating pumps, rotary pumps and special pumps.

New Words and Phases

pump [pʌmp] *n*. 泵
serve [səːv] *v*. 服务，供应
elevation [ˌeliˈveiʃən] *n*. 上升，提高
velocity [viˈlɔsiti] *n*. 速度，速率
vacuum [ˈvækjuəm] *n*. 真空
compress [kəmˈpres] *vt*. 压缩
handle [ˈhændl] *vt*. 处理，操作
reciprocate [riˈsiprəkeit] *v*. 往复

rotary ['rəutəri] *adj*. 旋转的
centrifugal pump 离心泵

1. The chief use of pumps is to add ( )
   (A) velocity to the fluid.
   (B) pressure to the fluid.
   (C) energy to the fluid.
2. Which of the following is true? ( )
   (A) It is possible for a pump to work by itself.
   (B) Various kinds of pumping equipment are widely used in industry.
   (C) centrifugal pumps are not of great importance in industry.

Reading Material

Acids, Bases and Their Neutralization

Definition of acids

An acid is a substance which contains hydrogen which can be displaced by a metal and which reacts with water to make hydrogen ions as the only positive ions.

酸、碱及其中和

酸的定义

酸是一种物质，它含有氢，氢能被金属置换。酸跟水反应，只生成一种带正电荷的离子，就是氢离子。

Acids properties

Acids can change the color of indicators. They can react with water to make hydrogen ions and react with reactive metals which displace the hydrogen in the acid. They can also react with metal oxides, metal hydroxides, and metal carbonates to make salts, when this happens the acid is neutralized.

酸的性质

酸能改变指示剂的颜色，能跟水反应生成水合氢离子，跟活泼金属反应，金属置换出酸中的氢。酸也能和金属氧化物、金属氢氧化物和金属碳酸盐反应生成盐，反应时酸变成中性。

Strong acid and weak acids

Some acids react completely with water forming hydrogen ions. The solution made conducts strongly because it contains so many ions. These acids are called strong acids. Other acids react incompletely with water, and only make a few hydrogen ions. Because there are so few ions present in the solution, it conducts weakly. Acids like this are called weak acids. So strong acids are fully ionized in water and weak acids are only partially ionized in water.

强酸和弱酸

有些酸跟水完全反应生成氢离子，由于含有大量的离子，所以溶液导电性加强，这些酸叫强酸。有些酸跟水不完全反应，只生成少量氢离子。由于含有的离子少，所以溶液的导电性弱，像这种酸叫弱酸。所以，强酸在水里全部离子化，弱酸在水里部分离子化。

Definition of bases

Substances which neutralize

碱的定义

能中和酸的物质叫碱，碱是金

acids are called bases. Bases are the oxides, hydroxides, or carbonates of metals. Most bases are insoluble, but some dissolve in water.

属的氧化物、氢氧化物或碳酸盐。大多数碱是不溶于水的，但有些能溶于水。

Strong and weak alkalis

Some alkalis split up completely into ions. They are called strong alkalis. Other alkalis react incompletely with water making only a few hydroxide ions. They are called weak alkalis. The solution made from a soluble base has a soapy, slippery feel—most soap are alkaline. They can change the color of indicators.

Neutralization

When an acid and a base react together they neutralize each other. The acid loses its acidic properties and the base loses its basic properties. The products of a neutralization reaction are a salt and water. During these reactions, the hydrogen ions from the acid react with the hydroxide ions from the base to make water molecules. All the other ions are spectators and just stay in solution. Heat is produced because all neutralizations are exothermic.

Using neutralization

There are a number of important examples of neutralization.

1. Adding lime (calcium hydroxide) to soil

Plants need nitrogen, phosphorus, and potassium compounds from the soil to grow well. Most plants take up these elements better when the soil is alkaline. Lime neutralizes acids in the soil making it alkaline.

2. Reducing acid rain and its effects

The coal burnt in power stations contains sulphur as an impurity. When the coal is burnt, the sulphur is burnt too. This produces sulphur dioxide, which dissolves in rain forming acid rain. By passing the burnt gases through lime, they are neutralized. The product, calcium sulphate, is used as plaster. Many lakes have become acidic due to acid rain. All the fish in them die. Spraying powered lime on to the lake neutralizes the water so that fish can once again survive.

3. Neutralizing stomach acids

The human stomach contains hydrochloric acid so that the enzyme pepsin can begin digesting protein. Sometimes too much acid is made and begins to attack the stomach wall causing pain. The extra acid can be neutralized by taking an "ant-acid". These always contain a base such as sodium or magnesium carbonate.

Acids with two hydrogens

Some acids like sulphuric, H_2SO_4, and carbonic, H_2CO_3, have two hydrogens in them. These acids can be completely neutralized when both hydrogens are replaced or partly neutralized when only one hydrogen is replaced. The salt made by partial neutralization is called an acid salt.

Salts

Salts are substances in which the hydrogen of an acid has been replaced by a metal, e. g. NaCl or $MgSO_4$. Each acid can produce a family of salts.

e. g.

hydrochloric acid produces chlorides

nitric acid produces nitrates

sulphuric acid produces sulphates

ethanoic acid produces ethanoates

The first part of salt, the metal part, is determined by the base used to react with the acid. So each base also produces a family of salts.

e. g.

sodium bases like sodium hydroxide produce sodium salts

magnesium bases like magnesium oxide produce magnesium salts

强碱和弱碱

有些碱在水中全部变成离子，这些碱是强碱。有些碱跟水发生不完全反应，只能生成少量的氢氧根离子，这些碱叫弱碱。可溶性碱形成的溶液有一种像肥皂样滑腻的感觉，大多数肥皂是碱性的。碱能改变指示剂的颜色。

中和

当一种酸和一种碱反应时它们相互中和，酸失去酸的性质，而碱失去碱的性质。中和反应的产物是一种盐和水。在反应时，酸里的氢离子跟碱里的氢氧根离子反应生成水。所有的其他离子只是旁观者，仍在溶液里。因为所有的中和反应是放热反应，所以有热量产生。

中和反应的例子

有许多中和反应的重要例子。

1. 在泥土里加石灰（氢氧化钙）

植物需要从土壤里吸取氮、磷和钾的化合物才能生长良好。当土壤呈碱性时，大多数植物能较好地吸取这些元素，石灰能中和土壤里的酸，使土壤变为碱性。

2. 减少酸雨和它的影响

发电厂里燃烧的煤炭含有杂质硫，当煤炭燃烧时，硫也燃烧，产生二氧化硫，它溶解在雨水里变成酸雨。如果让燃烧的气体通过石灰层，气体就被中和，产物是硫酸钙，用作石膏。许多湖泊由于酸雨而呈酸性，湖里的鱼都死了。向湖里撒石灰粉能中和湖水，鱼能重新获得生机。

3. 中和胃酸

人的胃里有盐酸，这样胃蛋白酶就能消化蛋白质，有时胃里会形成过多的胃酸，并开始损伤胃壁引起胃痛。这时服用一种抗酸剂就能中和过多的胃酸。它们是一种像碳酸钠或碳酸镁那样的碱。

有两个氢原子的酸

有些酸像硫酸（H_2SO_4）、碳酸（H_2CO_3）有两个氢原子，这些酸中当两个氢原子都被置换时，能全部中和，当只有一个氢原子被置换时，能部分被中和。因部分中和而形成的盐叫酸式盐。

盐

盐是酸里的氢被一种金属取代而形成的物质。比如：氯化钠或硫酸镁。每一种酸可以形成一族盐。

比如：

盐酸生成氯化物

硝酸生成硝酸盐

硫酸生成硫酸盐

醋酸生成醋酸盐

盐的前面部分，即金属部分是由用来跟酸反应的碱决定的。所以，每一种碱也能形成一族盐。

比如：

氢氧化钠生成钠盐

氧化镁生成镁盐

Unit 19

The Types of Titration

There are four types of titration methods.

(1) Acid-base titrations

Many compounds are either acids or bases. They can be titrated with a strong base or a strong acid. The end points of these titrations are easy to detect by means of an indicator.

(2) Precipitation titrations

The titrant forms an insoluble product with the analyte. For example, we can use silver nitrate solution to titrate with chlorine ion. Again, indicators can be used to detect the end points.

(3) Complexometric titrations

In complexometric titrations, the titrant is complexing agent. It can forms an water soluble complex with the analyte, a metal ion. EDTA is one of the most useful complexing agents. In the titration, it can react with many metal ions.

(4) Oxidation-Reduction titrations

In these titrations, an oxidizing agent reacts with a reducing agent, or vice versa. An oxidizing agent gains electrons and a reducing agent loses electrons in a reaction between them. Indicator can be used to detect the end point.

New Words

titrate [ˈtaitreit] *v*. 滴定，用滴定法测量

complexometric [ˌkɔmpleksˈɔmitrik] *adj*. 配位的

complexing [kəmˈpleksiŋ] *n*. 配位

electron [iˈlektrɔn] *n*. 电子

Expressions and Technical Terms

acid-base titration 酸碱滴定

strong acid 强碱

strong base 强酸

end point 终点

by means of 依靠，借助于

precipitation titration 沉淀滴定

complexometric titration 配位滴定

silver nitrate　硝酸银

complexing agent　配位剂

oxidation-reduction titration　氧化还原滴定

oxidizing agent　氧化剂

reducing agent　还原剂

vice versa　反之亦然

Exercises

A. Answer the following questions.

1. How many types of titration methods are there?

2. What can be used to detect the end points in an acid-base titration?

3. What does the titrant form with the analyte in precipitation titrations?

4. What is the titrant In complexometric titrations?

5. What does an oxidizing agent react with in a reduction-oxidation titrations?

B. Translate the following into English.

1. 酸碱滴定　　2. 配位滴定　　3. 氧化还原滴定

4. 沉淀滴定　　5. 终点　　6. 氧化剂

7. 还原剂　　8. 配位剂　　9. 强酸

10. 强碱

C. Reading comprehension. After reading a passage, choose the best answer to each question.

The movement of liquids and gases through pipelines is highly specialized and very important. Without the gas pipelines, the movement of large volumes of gases over great distances would be an economic impossibility. The liquid pipelines which used for either crude or refined petroleum, and for water, are an important part of the general transportation systems. Pipeline transportation has the following general characteristics: (a) fixed loading point and terminus, (b) high initial investment, and (c) the lowest cost for overland transportation.

The size of pipeline is governed by the amount of material to be transported through the line. The oil lines tend toward the smaller size, the gas toward the larger.

The pipeline, once installed, offers the cheapest method of transporting petroleum and natural gas, and certain other liquids and gases such as water, chlorine, hydrogen and steam.

New Words and Phases

movement [ˈmuːvmənt] *n*.运动，动作
pipeline [ˈpaipˌlain] *n*.管道，管线
specialize [ˈspeʃəlaiz] *vi*.专攻，专门研究
impossibility [imˌpɔsəˈbiləti] *n*.不可能之事，不可能
crude [kruːd] *adj*.天然的，未加工的
petroleum [piˈtrəuliən] *n*.石油
load [ləud] *n*.负荷，负载 *v*.装载
terminus [ˈtəːminəs] *n*.终点
initial [iˈniʃəl] *adj*.最初的，初始的
overland [ˌəuvəlænd] *adj*.陆路的，陆上的
govern [ˌgʌvən] *v*.统治，支配，管理
steam [stiːm] *n*.蒸汽，水汽
natural gas 天然气

1. Pipeline transportation has ( )
   (A) fixed loading point and terminus.
   (B) low initial investment.
   (C) lots of cost for overland transportation.
2. Pipeline transportation is often used for ( )
   (A) solid.
   (B) petroleum and natural gas, and certain other liquids and gases.

Reading Material

Iodometry—an Indirect Method

An important reactant in redox titrimetry is potassium iodide, KI. KI is a reducing agent that is useful in analyzing for oxidizing agents. The interesting aspect of the iodide is that it is most often used as an indirect method. This means that the oxidizing agent analyzed for is not measured directly by a titration with KI, but is measured indirectly by the titration of the iodide that forms in the reaction. The KI is actually added in excess, since it need not be measured at all. The experiment is called iodometry. Thus the percent of the oxidizing agent is calculated indirectly from the amount of titrant since the titrant actually reacts with I_2 and not "Oxidizing agent". This titrant is normally sodium thiosulfate ($Na_2S_2O_3$).

The sodium thiosulfate ($Na_2S_2O_3$) solution must be standardized. Several primary standard oxidizing agents are useful for this. Probably the most common one is potassium dichromate ($K_2Cr_2O_7$). Primary standard potassium bromate, $KBrO_3$, or potassium iodate, KIO_3, can also be used. Even primary standard iodine, I_2, can also be used. Usually in the standardization procedures, KI is again added to the substance to be titrated and the liberated iodine titrated with thiosulfate. If I_2 is the primary standard, it is titrated directly. The end points is usually detected with the use of a starch solution as the indicator. Starch, in the presence of iodine, is a deep blue color. When all of iodine is consumed by the thiosulfate, the color changes sharply from blue to violet providing a very satisfactory end point. Some important precautions concerning the starch, however, are to be considered. The starch solution should be fresh, should not be added until the end point is near, cannot be used in strong acid solutions, and cannot be used with solution temperatures above 40℃.

$$KI + \text{“O”} \longrightarrow \text{“R”} + I_2$$
$$I_2 + Na_2S_2O_3 \longrightarrow NaI + Na_2S_4O_6$$

An iodometry experiment. A solution of KI is added to a solution of the substance to be determined (represented by “O”). The reaction products are “R” and iodine, I_2. The amount of I_2, which is proportional to the amount of “O”, is then titrated with thiosulfate, $Na_2S_2O_3$.

碘量滴定法——一种间接的方法

氧化还原滴定法中一种重要的反应物是碘化钾。碘化钾在分析氧化剂时是一种有用的还原剂。碘化物有趣的一面是它最常被用在间接方法中。这意味着被分析的氧化剂不直接用碘化钾滴定来测定，而是通过滴定反应中形成的碘来间接测定。由于碘化钾不需被测量，因而实际上被过量加入。这一实验被称为碘量滴定法。由于滴定剂实际上是和碘而不是和氧化剂反应，因而，根据滴定剂的量，氧化剂的比例可被间接计算出。这种滴定剂通常是 $Na_2S_2O_3$。

硫代硫酸钠（$Na_2S_2O_3$）必须被标定。几种基准物的氧化剂被用于这一过程中。最常用的是重铬酸钾。基准物溴酸钾或碘酸钾也能用于标定，甚至基准物的碘也能用于标定。在标定程序中，碘化钾加入到被滴定的物质中，释放出的碘用硫代硫酸钠来滴定。如果碘是基准物，可直接被滴定。终点通常用淀粉溶液作指示剂来检测出。淀粉遇到碘呈深蓝色。当所有的碘被硫代硫酸盐消耗掉时，颜色明显由蓝变成紫色，提供了一个很明显的终点。然而，有关淀粉的一些重要预防措施需要考虑。淀粉溶液应是新鲜的，应到接近终点时才加入，不能用于强酸溶液中，不能用在 40℃以上的溶液中。

$$KI + \text{“O”} \longrightarrow \text{“R”} + I_2$$
$$I_2 + Na_2S_2O_3 \longrightarrow NaI + Na_2S_4O_6$$

碘量滴定法实验。碘化钾被加入到待测定溶液（用“O”来表示）中。反应产物是碘和用“R”表示的还原物。与“O”成比例的碘的量，可用硫代硫酸钠来滴定。

Unit 20

Gas Chromatography

In analytical chemistry, gas chromatography is a technique for separating chemical substances. Because of its simplicity, sensitivity, and effectiveness, gas chromatography is one of the most important tools in chemistry. It is widely used for purification of compounds and for the determination of some chemical constants and etc..

In the gas chromatography, we first introduce the test mixture or sample into an inert gas, the carrier stream. Liquid samples should be vaporized before injection into the carrier stream. Then the gas stream is passed through the packed column in which there is a stationery phase. The components of the sample having the greater interaction with the stationery phase are retarded to a greater extent. The components that have a lower interaction travel through the column at a faster rate. So sample components separate from each other.

As each component leaves the column, it passes through a detector and then goes to a collector.

New Words

chromatography [ˌkrəuməˈtɔgrəfi] *n*. 色谱法
simplicity [simˈplisiti] *n*. 简单，简易
determination [ditəːmiˈneiʃən] *n*. 测量，测定
sample [ˈsæmpl] *n*. 样品
injection [inˈdʒekʃən] *n*. 注射
column [ˈkɔləm] *n*. 圆柱，柱状物
stationery [ˈsteiʃənəri] *adj*. 固定的
component [kəmˈpəunənt] *n*. 成分
retard [riˈtaːd] *vt*. 延迟，阻碍
detector [diˈtektə] *n*. 探测器
collector [kəˈlektə] *n*. 收集器

Expressions and Technical Terms

gas chromatography 气相色谱
analytical chemistry 分析化学
inert gas 惰性气体
carrier stream 载体流
packed column 填料柱

stationery phase　固定相

Exercises

A. Answer the following questions.

1. What are the uses of the gas chromatography?

2. What are the uses of the inert gas in the chromatography?

3. Should the liquid samples be vaporized before injection into the carrier stream?

4. Why can sample components separate from each other?

B. Translate the following into English.

1. 气相色谱　　2. 惰性气体　　3. 收集器
4. 检测器　　5. 填料塔　　6. 固定相
7. 纯化

C. Reading comprehension. After reading a passage, choose the best answer to each question.

The molecules of a substance are in continual motion. Imagine a crowd of people leaving a football match. They are all trying to get away as quickly as possible, and this is exactly how the molecules of a gas behave. They try to spread out, or in other words, to diffuse.

Diffusion can be described by opening a bottle of perfume in a room. The molecules of the perfume are in the bottle. Immediately the bottle is opened, they start to diffuse into the air, and the perfume can be smelt at the other end of the room. The longer the bottle is left open, the stronger the smell becomes.

Finally, when there is an equal concentration of molecules inside and outside the bottle, diffusion ceases. According to the principles of diffusion, the process continues until the concentration of perfume molecules is equal in all parts of the room.

New Words

motion [ˈməuʃən] *n*. 运动，动作
crowd [kraud] *n*. 人群，群众
spread [spred] *v*. 伸展，展开
diffuse [diˈfjuːz] *v*. 散播，传播，散开
diffusion [diˈfjuːʒən] *n*. 扩散，传播，漫散
cease [siːs] *v*. 停止
principle [ˈprinsəpl] *v*. 原则，原理

1. The molecules of a substance (　　)

(A) do not move.

(B) are in continual motion.

(C) move as slowly as possible.

2. The diffusion of the perfume molecules ceases, when (　　)

(A) the concentration in the bottle is lower than that in the room.

(B) the concentration of the perfume molecules is equal in all parts of the room.

Reading Material

Liquid Chromatography

For liquid chromatography, the procedure can be performed either in a column or on a plane. Columnar liquid chromatography is used for qualitative and quantitative analysis in a manner similar to the way in which to the way in which gas chromatography is employed. For chemical analysis the most popular category of columnar liquid chromatography is high performance liquid chromatography (HPLC). The method uses a pump to force one or more mobile phase solvents through high efficiency, tightly packed columns. As with gas chromatography, an injection system is used to insert the sample into the entrance to the column, and a detector at the end of the column monitors the separated analyte components.

The stationery phase that is used for plane chromatography is held in place on a plane. Typically the stationery phase is attached to a plastics, metallic, or glass plate. Occasionally, a sheet of high quality filter paper is used as the stationery phase. The sample is added as a spot or a thin strip at one end of the plane. The mobile phase flows over the spot by capillary action during ascending development or as a result of the force gravity during descending development. During ascending development, the end of the plane near and below the sample is dipped into the mobile phase, and the mobile phase moves up and through the spot. During descending development, the mobile phase is added to the top of the plane and flows downward through the spot.

Qualitative analysis is performed by comparing the retardation factor of the analyte components with the retardation factors of known substances. The retardation factor is defined as the distance from the original sample spot that the component has moved divided by the distance that the mobile phase front has moved and is constant for a solute in a given solvent. Quantitative analysis is performed by measuring the sizes of the developed spots, by measuring some physical property of the spots, or by moving the spots from the plane and assaying them by another procedure.

液相色谱

对于液相色谱而言，该程序可在柱子中或平板上开展。柱式液相色谱用于定性或定量分析，与气相色谱中使用的方法相似。对于化学分析来说，最普遍的柱式液相色谱是高效液相色谱。该方法是使用泵使一种或多种流动相溶剂穿过高效的、紧密的填料柱。像气相色谱一样，使用注射系统将样品注入填料柱的入口，在柱子的末端用检测器检测被分离的组分。

薄层色谱中的固定相被固定在平板上。通常固定相被附着到塑料、金属或玻璃的平板上。偶尔，一层高质量的滤纸也被用作固定相。样品以点滴或细线的形式被加到平板的一端。流动相在上行展开法中通过毛细作用流过加样点；流动相在下行展开法中由于重力的作用流过加样点。在上行展开法中，接近且在加样点下方的平板一端被浸入到流动相中，流动相向上移动通过加样点。在下行展开法中，移动相被加入到平板的顶端向下流动并通过加样点。

通过比较分析物组分和已知物质的保留因子来进行定性分析。保留因子定义为组分移动的距离除以移动相前缘移动的距离，对于给定溶剂中的溶质而言，该数值是恒定的。通过测量展开点的大小或测量斑点的物理性质来进行定量分析，或者将斑点从平板上移动下来，用其他程序对其进行试验来进行定量分析。

Unit 21

The Nature of Biotechnology

Biotechnology is an area of applied bioscience and technology which involves the practical application of biological organisms, or their subcellular components. Biotechnology utilizes bacteria, yeasts, fungi, algae, plant cells, or cultured mammalian cells as constituents of industrial processes.

Biotechnological processes will normally involve the production of cells or biomass, and the achievement of desired chemical transformations. The latter may be further subdivided into the formation of a desired end product (enzymes, antibiotics, organic acids) and decomposition of a given starting material.

The reactions of biotechnological processes can be catabolic, in which complex compounds are broken down to simpler ones, or anabolic whereby simpler molecules are built up into more complex ones.

Biotechnology includes fermentation processes ranging from beers and wines to bread, cheese, antibiotics and vaccines, water and waste treatment, parts of food technology.

New Words

biotechnology [ˌbaiətekˈnɒlədʒi] *n*. 生物技术，生物工艺

subcellular [sʌbˈseljʊlə (r)] *adj*. 亚细胞的

bacteria [bækˈtiəriə] *n*. 细菌

yeast [jiːst] *n*. 酵母

fungi [ˈfʌŋgiː] *n*. 真菌，霉菌；菌类

algae [ˈældʒiː] *n*. 藻类；海藻

mammalian [məˈmeiliən] *adj*. 哺乳类动物的

enzyme [ˈenzaim] *n*. 酶

antibiotics [ˌæntibaiˈɒtiks] *n*. 抗生素；抗生学

catabolic [ˌkætəˈbɒlik] *adj*. 分解代谢的；异化的

anabolic [ˌænəˈbɒlik] *adj*. 合成代谢的；同化的

vaccine [ˈvæksiːn] *n*. 疫苗

Expressions and Technical Terms

biotechnological process 生物技术过程

chemical transformation　化学转变；化学变化

end product　最终产品；最后结果

starting material　起始物料，原始材料；原材料

fermentation process　发酵过程；发酵工艺

Exercises

A. Answer the following questions.

1. What is the meaning of biochemistry?

2. Biotechnology utilizes (　　).

(A) bacteria and yeasts

(B) fungi and algae

(C) plant cells or cultured mammalian cells

(D) all of above

3. what do biotechnological processes normally involve?

4. which of the following includes fermentation processes?

(A) beers and wines

(B) bread and cheese

(C) antibiotics and vaccines

(D) all of above

B. Translate the following into English.

1. 生物技术　　2. 细菌　　3. 酵母

4. 真菌　　5. 藻类　　6. 酶
7. 抗生素　　8. 疫苗　　9. 分解代谢
10. 合成代谢　　11. 发酵过程

C. Reading comprehension. After reading a passage, choose the best answer to each question.

The specific processes are catalysed by microorganisms, plant or animal cells, or products derived from them such as enzymes. The organisms of biotechnology can be harvested for biomass, can be used to perform chemical conversions, and may be the source of biologically active molecules, including enzymes and monoclonal antibodies.

Gene manipulation techniques have brought a new dimension to applied genetics and have created the potential for completely novel industrial processes, for example, human interferon produced by bacterial cells. Significant developments are also occurring in process and control engineering and fermentation technology which will further advance the development of biologically-based industrial activity.

Biotechnology appears to be an area of expansion and opportunity involving many sectors of industry, including agriculture, food and feedstuffs, pharmaceutical, energy and water industries. It will play a major role in the production of new drugs, hormones, vaccines and antibiotics, cheaper and more reliable supplies of energy and (in the longer term) chemical feedstuffs, improved environmental control and waste management. Biotechnology will be largely based on renewable and recyclable materials thus being better fitted to the needs of a world where energy will become increasingly more expensive and in short supply.

New Words and Phases

catalyse [ˈkætəlaiz] *vt*. 催化；促成

microorganism [maikrəʊˈɔːgənizəm] *n*.［微］微生物

biomass [ˈbaiəʊmæs] *n*. 生物量

monoclonal antibody 单克隆抗体

gene manipulation technique 基因操作技术

interferon [ˌintəˈfiərɒn] *n*.［生化］干扰素

feedstuff [ˈfiːdstʌf] *n*. 饲料，原料

hormone [ˈhɔːməʊn] *n*.［生理］激素，荷尔蒙

vaccine [ˈvæksiːn] *n*. 疫苗

1. The specific processes of biotechnology are not catalysed by ( ).

(A) microorganisms

(B) plant or animal cells

(C) chemical enzymes

(D) enzymes derived from plant or animal cells

2. ( ) have brought a new dimension to applied genetics and have created the potential for completely novel industrial processes.

(A) Human interferon produced by bacterial cells

(B) Gene manipulation techniques

(C) Fermentation technology

(D) Control engineering

3. Biotechnology appears to be an area of expansion and opportunity involving many sectors of industry, including ( ).

(A) agriculture

(B) food and feedstuffs

(C) pharmaceutical, energy and water industries

(D) all of above

Reading Material

Historical Evolution of Biotechnology

生物技术的历史演变

Contrary to popular belief, biotechnology is not a new pursuit but in reality dates far back into history. In practice, four major developmental phases can be identified, arriving at modern biotechnological systems.

与普遍的看法相反，生物技术并不是一项新的追求，实际上它可以追溯到很久以前。在实践中，可以确定四个主要发展阶段，达到现代生物技术系统。

Other microbially based processes such as the production of fermented milk products, e. g. cheeses and yogurts, and various oriental foods, e. g. soy sauce, tempeh etc., can equally claim distant ancestry. The most recent introduction is mushroom cultivation which probably dates back many hundreds of years for Japanese cultivation.

其他以微生物为基础的工艺，如发酵牛奶产品的生产，如奶酪和酸奶，以及各种东方食品，如酱油、豆豉等，都可以声称具有遥远的祖先。最近引进的是蘑菇栽培，这可能可以追溯到几百年前的日本栽培。

Activities such as baking, brewing and wine making are known to date back several millenias; the ancient Sumarians and Babylonians were drinking beer by 6000 B. C.,

烘焙、酿造和酿酒等活动可以追溯到几千年前；公元前6000年，古苏美尔人和巴比伦人已经开始喝啤酒了；公元前4000年，埃及人已经开始烤有酵的面包了；

the Egyptians were baking leavened bread by 4000 B. C., while wine was known in the Near East by the time of the book of Genesis. The recognition that these processes were being affected by living organisms, yeasts, was not formulated until the 17th century. Definitive proof of the fermentative abilities of these minute organisms came from the seminal studies of Pasteur between 1857 and 1876. Pasteur can justifiably be considered as the father of biotechnology.

而在《创世纪》所描述的时代，近东人已经知道酒了。直到17世纪，人们才认识到这些过程受到活的有机体——酵母的影响。这些微小生物发酵能力的确切证据来自1857年至1876年巴斯德的开创性研究。巴斯德有理由被认为是生物技术之父。

It cannot be ascertained whether these microbial processes arose by accidental observation or by intuitive experimentation, but their further and continued development were early examples of man's abilities to use the vital activities of organisms for his own needs. In more recent times, just as these processes have become more reliant on advanced technology, their contribution to the world economy has equally increased out of all proportion to their humble origins.

这些微生物的反应过程是偶然观察引起的还是由实验引起的，尚不能确定，但它们进一步和持续的发展是人类利用生物体满足自己需要的能力的早期例证。近年来，这些过程变得更加依赖先进技术，它们对世界经济的贡献也同样超出了它们卑微的起源。

Unit 22

Fermentation Technology

The origins of fermentation technology were large with the use of microorganisms for the production of foods and beverages such as cheeses, yogurts, alcoholic beverages, vinegar, fermented pickles, sausages and soya sauce. The present-day production of these products is essentially scaled up.

Future markets are largely assured for fermentation products because it is not possible to produce them economically by other chemical means. Furthermore, economies in production will also occur by genetically engineering organisms. The commercial market for products of fermentation technology is almost unlimited but will ultimately depend on economics and safety considerations.

The processes of commercial fermentation are in essence very similar no matter what organism is selected, what medium is used and what product is formed. In all cases, large numbers of cells with uniform characteristics are grown under defined, controlled conditions. The same apparatus with minor modifications can be used to produce an enzyme, an antibiotic, an organic chemical or a single cell protein.

All biotechnological processes are carried out within a containment system or bioreactor. The physical form of most common bioreactors has not altered much over the past 30 years. Recently, however, many novel forms have been designed and they may play an increasingly active part in biotechnology. The main function of a bioreactor is to minimize the cost of producing a product or service.

New Words

fermentation [ˌfəːmenˈteiʃn] *n.* 发酵
beverage [ˈbevəridʒ] *n.* 饮料
yogurt [ˈjɒgət] *n.* 酸奶酪，酸乳
vinegar [ˈvinigə] *n.* 醋
pickle [ˈpikl] *n.* 泡菜；盐卤；腌制食品
sausage [ˈsɒsidʒ] *n.* 香肠；腊肠
medium [ˈmiːdiəm] *n.* 培养基

antibiotic [ˌæntibaiˈɒtik] *n*. 抗生素，抗菌素

bioreactor [ˈbaioriˌæktər] *n*. [细胞] 生物反应器

minimize [ˈminimaiz] *v*. 使减少到最低限度；贬低

Expressions and Technical Terms

fermentation technology 发酵工艺；发酵技术

soya sauce 酱油

scale up 向上扩展；按比例放大

a single cell protein 单细胞蛋白质

Exercises

A. Answer the following questions.

1. The origins of fermentation technology were large with the use of microorganisms for the production of ( ).

(A) cheeses and yogurts

(B) sausages and soya sauce

(C) alcoholic beverages and vinegar

(D) all of above

2. Why are future markets largely assured for fermentation products?

3. Which of the following is not true about the bio-

reactor?

(A) All biotechnological processes are carried out within a bioreactor.

(B) The physical form of most common bioreactors has not altered much over the past 30 years.

(C) The main function of a bioreactor is to minimize the cost of producing a product or service.

(D) many novel forms have not been designed.

B. Translate the following into English.

1. 发酵 2. 饮料 3. 醋

4. 泡菜 5. 培养基 6. 抗生素

7. 生物反应器 8. 单细胞蛋白质

C. Reading comprehension. After reading a passage, choose the best answer to each question.

The bioreactor is the containment system for the biological reactions of a biotechnological process. It will provide the correct environment for optimization of organism growth and metabolic activity. it should prevent contamination of the production culture from the environment, while also preventing the release of the culture into the environment.

Many bioreactor systems will be required to operate under aseptic conditions. In most systems of industrial fermentation, pure cultures of the organism will be used and the presence of foreign unwanted contaminating organisms can influence the process in several ways, for example, by interfering with the catalyst, destroying the product, producing substances that can impair downstream processing, and introducing toxic molecules into the systems.

To avoid these problems the medium, bioreactor and all ancillary pipework must be sterilized and all incoming air must be passed through sterilized glass wool to remove contaminants.

New Words and Phases

optimization [ɒptimaiˈzeiʃən] *n*. 优化；最优化
metabolic [ˌmetəˈbɒlik] *adj*. 新陈代谢的
culture [ˈkʌltʃər] *n*. 培养，培养物
downstream processing 下游加工过程
ancillary pipework 辅助管道工程
sterilized [ˈsterəlaizd] *adj*. 无菌的；已消过毒的
glass wool 玻璃棉

1. Which of the following is not true?

(A) The bioreactor is the containment system for biological reactions.

(B) The bioreactor provides the correct environment for the optimization of organism growth and metabolic activity.

(C) The bioreactor can not prevent the release of the culture into the environment.

(D) The bioreactor should prevent contamination of the production culture from the environment.

2. The presence of foreign unwanted contaminating organisms can influence the biotechnological process in several ways, for example, by (　　).

(A) interfering with the catalyst

(B) destroying the product

(C) producing substances that can impair downstream processing

(D) All of the above

3. In order to operate under aseptic conditions, (　　).

(A) all incoming air must not be removed contaminants

(B) The medium must be sterilized

(C) The bioreactor must not be sterilized

(D) all ancillary pipework must not be sterilized

Reading Material

Media Design

Media are designed to meet the nutritional demands of the producer organism, the objectives of the process and the scale of the operation. For most large scale biotechnological processes cost, availability and handling properties of the medium components are major factors in determining selection.

The basic nutritional requirements of heterotrophic microorganisms are an energy or carbon source, an available nitrogen source, inorganic elements and, for some microorganisms, growth factors. For most biotechnological processes carbon and nitrogen sources are more often derived from relatively complex mixtures of cheap natural products or byproducts, while trace metals are normally present in sufficient amounts in the tap water or in the main raw materials. Salts are often added as

培养基设计

培养基的设计是为了满足生产者有机体的营养需求、生产加工目标和操作的规模。对于大多数大规模的生物技术过程，成本、介质成分易获取性和易处理性是选择时考虑的主要因素。

异养微生物的基本营养需求是能量或碳源、可用氮源、无机元素和某些微生物的生长因子。对于大多数生物技术过程来说，碳和氮来源往往来自于相对复杂的廉价天然产品或副产品的混合物，而微量金属通常在自来水或主要原材料中有足够量的存在。盐常被添加作为氮、磷、硫或钙的补充来源。在需要时，生长因子可以以纯物质的形式提供，但出于经济原因，一般以植物或动物提取物的形式添加。

supplementary sources of nitrogen, phosphorus, sulfur or calcium. Growth factors when required may be supplied in pure form but for economic reasons would generally be added as a plant or animal extract.

The main types of growth factors required are B-group vitamins or related compounds, certain amino acids and some fatty acids. A proper balance of carbon and nitrogen sources can be important to the pH pattern of processes if pH control is not applied. For most processes, the nutrients have to be dissolved in water. In batch systems, nutrients are normally all present in the initial volume. Fermentation reactions can be further regulated and controlled in batch cultures by feeding some of the nutrients on a specific rate basis. In this way, essential inducer concentrations can be maintained. Nutrient availability will exert strong physiological control over fermentation reactions and product formation.

所需生长因子的主要类型是B族维生素或相关化合物、某些氨基酸和某些脂肪酸。如果不进行pH控制，适当的碳和氮源平衡可能对过程的pH很重要。在大多数加工过程中，营养物必须溶解在水中。在间歇生产系统中，营养物质通常都在初始物料中。在分批培养中，可以通过按特定的速率添加某些营养物质来进一步调节和控制发酵反应。这样就可以维持必要的刺激物浓度。营养利用率将对发酵反应和产物形成产生强有力的生理学上的控制。

Unit 23

Is It a Cosmetic, a Drug, or Both?

The legal difference between a cosmetic and a drug is determined by a product's use. Different laws apply to each type of product. Firms sometimes violate the law by marketing a cosmetic with a drug claim, or by marketing a drug as if it were a cosmetic, without adhering to requirements for drugs.

How does the law define a cosmetic? Cosmetics are defined as articles intended to be rubbed, poured, sprinkled, or sprayed on the human body for cleaning, beautifying, or promoting attractiveness. For example, perfumes, lipsticks, shampoos, hair colors and toothpastes are included in this definition.

How does the law define a drug? Drugs are defined as articles intended for use in the diagnosis, cure, mitigation, treatment, or prevention of disease. The articles intended to affect the structure or any function of the body of man or other animals are also drugs.

How can a product be both a cosmetic and a drug? Some products meet the definitions of both cosmetics and drugs. This may happen when a product has two intended uses. For example, shampoo is a cosmetic because its intended use is to clean the hair. An antidandruff treatment is a drug because its intended use is to treat dandruff. Consequently, an antidandruff shampoo is both a cosmetic and a drug.

New Words

cosmetic [kɒzˈmetik] *adj*. 化妆用的，美容的；*n*. 化妆品，美容品

legal [ˈliːgl] *adj*. 法律的；合法的；法定的

violate [ˈvaiəleit] *vt*. 违反；侵犯，妨碍；亵渎

pour [pɔːr] *vi*. 灌，倒；倾泻

sprinkle [ˈspriŋkl] *vi*. 撒，洒；用……点缀

spray [sprei] *n*. 喷雾，喷雾剂；*vt*. 喷射

perfume [ˈpɜːfjuːm] *n*. 香水；芳香，香味

lipstick [ˈlipstik] *n*. 口红，唇膏

shampoo [ʃæmˈpuː] *n*. 洗发；洗发精

diagnosis [ˌdaiəgˈnəʊsis] *n*. 诊断，判断

mitigation [ˌmitiˈgeiʃn] *n*. 减轻；缓和；平静

dandruff [ˈdændrʌf] *n*. 头皮屑

Expressions and Technical Terms

adhere to 坚持；黏附；拥护

be defined as… 规定（被称为）……

intend to… 目的在于；用来……

Exercises

A. Answer the following questions.

1. What determine the legal difference between a cosmetic and a drug?

2. How does the law define a cosmetic?

3. About a drug，which one is untrue?

（A） The drug can affect the structure of the body of a man.

（B） The drug can be used to clean human body.

（C） The drug can be used to prevent a disease.

（D） The drug is impossible to be a cosmetic.

B. Translate the following into English.

1. 化妆品　2. 药品　3. 香水

4. 洗发精　5. 口红　6. 头皮屑

7. 牙膏

C. Reading comprehension. After reading a passage, choose the best answer to each question.

How is a product's intended use established? Intended use may be established in a number of ways. Among them are:

Claims are stated on the product labeling, in advertising, on the internet, or in other promotional materials. Certain claims may cause a product to be considered a drug, even if the product is marketed as if it were a cosmetic. Such claims establish the product as a drug because the intended use is to treat or prevent disease or otherwise affect the structure or functions of the human body. Some examples are claims that products will restore hair growth, reduce cellulite, treat varicose veins, or revitalize cells.

Consumer perception may be established through the product's reputation. This means asking why the consumer is buying it and what the consumer expects it to do.

Ingredients that may cause a product to be considered a drug because they have a well known (to the public and industry) therapeutic use. An example is fluoride in toothpaste.

New Words and Phases

claims [kleimz] *n.* 要求，请求权
labeling ['leibliŋ] *n.* 标签；标记
promotional material 宣传资料；宣传片
restore [ri'stɔː] *vt.* 恢复；修复；归还
cellulite ['seljʊlait] *n.* 脂肪团
varicose veins 静脉曲张
revitalize [riː'vaitəlaiz] *vt.* 使……复活；使……复兴
ingredient [in'griːdiənt] *n.* 成分，原料；要素，因素
therapeutic [θerə'pjuːtik] *adj.* 治疗的；治疗学的

1. Which of the following claims may not cause a product to be considered a drug?

(A) to treat disease

(B) to prevent disease

(C) to affect the structure or functions of the human body

(D) to clean the hair

2. Product intended use may be established in a number of ways except ( ).

(A) state the definition of a drug or a cosmetic

(B) state certain claims to cause a cosmetic to be considered a drug

(C) ask the consumer about the reason and expectation

(D) state the component, which has a well-known especial use

Reading Material

Soap

肥皂

How FDA defines "soap"

Not every product marketed as soap meets the FDA's definition of the term. FDA interprets the term "soap" to apply only when the bulk of the nonvolatile matter in the product consists of an alkali salt of fatty acids and the product's detergent properties are due to the alkali-fatty acid compounds, and the product is labeled, sold, and represented solely as soap.

If a product intended to cleanse the human body does not

FDA 如何定义"肥皂"

并不是每一种作为肥皂销售的产品都符合 FDA 的定义。FDA 对"肥皂"一词的解释仅适用于以下情况：产品中的大部分非挥发性物质由脂肪酸的碱盐组成，且产品的洗涤性能是由碱性脂肪酸化合物决定的，且产品标签、销售和表达仅为肥皂。

如果一种用于清洁人体的产品不符合上述香皂的所有标准，

meet all the criteria for soap，as listed above，it is either a cosmetic or a drug. For example：if a product consists of detergents or primarily of alkali salts of fatty acids and is intended not only for cleansing but also for other cosmetic uses，such as beautifying or moisturizing，it is regulated as a cosmetic.

那么它要么是化妆品，要么是药物。例如：如果一种产品由洗涤剂或主要由脂肪酸的碱盐组成，而且不仅用于清洁，还用于其他化妆品用途，如美容或保湿，那么它就被规定为化妆品。

If a product consists of detergents or primarily of alkali salts of fatty acids and is intended not only for cleansing but also to cure，treat，or prevent disease or to affect the structure or any function of the human body，it is regulated as a drug.

如果一种产品由洗涤剂或主要是脂肪酸的碱盐组成，而且不仅用于清洁，还用于治疗或预防疾病，或影响人体的结构或任何功能，则该产品被规定为药物。

If a product is intended solely for cleansing the human body and has the characteristics consumers generally associate with soap，does not consist primarily of alkali salts of fatty acids，it may be identified in labeling as soap，but it is regulated as a cosmetic.

如果一种产品仅用于清洁人体，且具有消费者通常与肥皂联系在一起的特征，主要不包括脂肪酸的碱盐，则可以在标签中标识为肥皂，但它被规定为化妆品。

Unit 24

Emulsion

Paints, polishes, metal cutting oils, ice cream, and cosmetics, etc. are all emulsions or are used in emulsified form.

An emulsion is a significantly stable suspension of particles of a liquid. The term "significantly stable" means relative to the intended use and may range from a few minutes to a few years. Two types of emulsions are recognized: a) macroemulsions. this is the usual type of emulsions in which the particles range from 0.2～50μm in size; b) microemulsions. This type of emulsions has particles from 0.01～0.2μm in size.

If the diameter of the dispersed particles is about 1μm, the emulsion is milky white; 1～0.1μm, blue-white; 0.1～0.05μm, gray and semitransparent; less than 0.05μm, transparent. Thus macroemulsions are opaque and microemulsions are either transparent or semitransparent to visible light.

Based on the nature of the dispersed phase, emulsions can also be classified into two types: oil-in-water (o/w), and water-in-oil (w/o). The o/w type is a dispersion of oil in an aqueous phase, and the oil is called the discontinuous phase. On the other hand, the w/o type is a dispersion of water or aqueous solution in an oily liquid. In this case, the water is called the discontinuous phase (the inner phase), the oil is called the continuous phase (the outer phase).

New Words

paint [peint] *n*. 涂料；颜料
polish ['pɒliʃ] *n*. 磨光，擦亮；擦亮剂
emulsion [i'mʌlʃn] *n*. 乳剂；乳状液
macroemulsion ['mækrəʊ'mʌlʃn] *n*. 粗乳液
microemulsion ['maikrəʊ'mʌlʃn] *n*. 微乳液
transparent [træns'pærən] *adj*. 透明的，清澈的
semitransparent [ˌsemitræns'pærənt] *adj*. 半透明的
disperse [di'spɜːs] *v*. 分散，散布

Expressions and Technical Terms

oil-in-water (o/w) 水包油

water-in-oil（w/o）油包水
aqueous phase 水相
discontinuous phase 不连续相

Exercises

A. Answer the following questions.

1. According to the text of the definition of emulsion，the emulsion may not include（ ）?

（A）paints （B）polishes

（C）pesticides （D）alcohol

2. What is the difference between macromulsions and microemulsions?

3. What types can emulsions be classified according to the nature of the dispersed phase?

B. Translate the following into English.

1. 擦亮剂 2. 乳状液 3. 粗乳液

4. 微乳液 5. 水包油 6. 油包水

7. 水相 8. 不连续相

C. Reading comprehension. After reading a passage，choose the best answer to each question.

Essential oils are subtle，therapeutic-grade oils distilled from plants，shrubs，flowers，trees，roots， bushes and seeds. They are oxygenating and help transport nutrients to the cells of our body. Without oxygen, nutrients cannot be assimilated; therefore, the oxygenating essential oils can help us maintain our health.

One of the most popular applications for essential oils is in the field of aromatherapy. This highly skilled art uses essential oils to assist in the healing of physical, psychological and aesthetic ailments. It is the only therapy that utilizes the most neglected of the senses, smell.

There are many ways to incorporate the benefits of aromatherapy into our daily lives. They may be used to stimulate and invigorate us in the morning, and then to calm and restore our peace of mind at the end of the day. Essential oils may soothe inflammation, act as an antiseptic, help dull pain and stimulate digestion.

New Words and Phases

essential oil　精油

therapeutic-grade　治疗级别的

distill [di'stil]　*vt*. 提取；蒸馏

oxygenate ['ɒksidʒəneit]　*v*.（使）富氧；氧化

assimilate [ə'siməleit] *vi*. 吸收；同化
aromatherapy [əˌrəʊmə'θerəpɪ] *n*. 芳香疗法
psychological [ˌsaikə'lɒdʒikl] *adj*. 心理的，精神的
aesthetic [iːs'θetik] *adj*. 美的；美学的
stimulate ['stimjuleit] *vt*. 刺激；鼓舞，激励
invigorate [in'vigəreit] *vt*. 鼓舞；使精力充沛
restore [ri'stɔː] *vt*. 恢复；还原
inflammation [ˌinflə'meiʃn] *n*. 炎症
antiseptic [ˌænti'septik] *adj*. 防腐的，抗菌的
digestion [dai'dʒestʃən] *n*. 消化；消化能力

1. Which of the following claims is not true about the essential oils?

(A) Essential oils are subtle, therapeautic-grade oils distilled from plants, shrubs, flowers, trees, roots, bushes and seeds.

(B) Essential oils are hydrogenating.

(C) The oxygenating essential oils can help us maintain our health.

(D) Essential oils help transport nutrients to the cells of our body.

2. According to the passage, one of the most popular applications for essential oils is (　　).

(A) in the field of aromatherapy

(B) in the healing of physical ailments

(C) maintaining our health

(D) in the healing of psychological ailments

3. There are many ways to incorporate the benefits of aromatherapy into our daily lives. Essentials oils may ( ).

(A) soothe inflammation

(B) act as an antiseptic

(C) help dull pain and stimulate digestion

(D) all of the above

Reading Material

the Stability of Emulsions

In the formation of emulsions, the fine dispersion of the inner phases produces a tremendous increase in the area of the interface between the two phases, these results in turn in a correspondingly large increase in the interfacial free energy of the system. Thus the emulsion produced is highly unstable. The function of the emulsifier is to stabilize

乳液的稳定性

在乳化液的形成过程中，内部相的精细分散使两相之间的界面面积大大增加，从而导致系统的界面自由能相应地大幅度增加。因此，所产生的乳液极不稳定。乳化剂的作用是使这个不稳定的体系稳定足够的时间，使乳化剂能够履行它的功能。

this unstable system for a sufficient time so that the emulsion can perform its function.

The stabilization the emulsifier does is as follows: the emulsifier molecules adsorb at the L/L interface as an oriented interfacial film; reduce the interfacial tension markedly; decrease the rate of coalescence of the dispersed liquid particles by forming mechanical and/or electrical barriers around them.

乳化剂的稳定作用如下：乳化剂分子吸附在L/L界面上形成定向界面膜；显著降低界面张力；通过在分散的液体粒子周围形成机械和/或电子屏障，降低其聚结率。

Factors that determine the stability of an emulsion include physical nature of the interfacial film resulting from the adsorbed surfactants, existence of an electrical or steric barrier to coalescence on the dispersed droplets, size distribution of droplets, phase volume ratio, temperature, and so on. Anything that disturbs the interface decreases the stability of the emulsion, the increased vapor pressure resulting from the increase in temperature causes an increased flow of molecules through the interface, and decrease the stability of the emulsion.

决定乳液稳定性的因素包括吸附表面活性剂所形成的界面膜的物理性质、分散液滴上是否存在电阻或位阻、液滴的大小分布、相体积比、温度等。任何扰乱界面的东西都会降低乳化液的稳定性，温度升高引起的蒸气压增加会导致通过界面的分子流动增加，从而降低乳化液的稳定性。

Unit 25

Preservatives

More and more products contain ingredients that make cosmetic systems very difficult to preserve. Preservation is mandated by the Food and Drug Administration. The products that are sold to the general public must be safe for use when it is applied to the body and they must be free of contamination.

Cosmetics must be preserved adequately to kill the microorganisms that are introduced by the consumers themselves. But too much preservative in cosmetics is no good either. A microbiology challenge test will show the minimum amount of preservative needed to provide a kill in a product.

As a formulator, one must do more research to find the right combination of preservatives to kill any microorganisms. More research is required because many preservatives that worked very well in the past are not being used any longer. This could be due to safety issues.

New Words

preservative [priˈzɜːvətiv] *n*. 防腐剂，保护剂
mandate [ˈmændeit] *v*. 强制执行，颁布
administration [ədˌminiˈstreiʃn] *n*. 管理，行政；管理部门，行政部门
contamination [kənˌtæmiˈneiʃn] *n*. 污染
microorganism [ˌmaikrəʊˈɔːgənizəm] *n*. 微生物
microbiology [ˌmaikrəʊbaiˈɒlədʒi] *n*. 微生物学
formulator [ˈfɔːmjuleitə] *n*. 配方设计师
combination [ˌkɒmbiˈneiʃn] *n*. 组合，结合，联合

Expressions and Technical Terms

Food and Drug Administration　食品和药物管理局
be free of…　远离……；免于……
microbiology challenge test　微生物挑战试验

Exercises

A. Answer the following questions.

1. Which one mandates the preservation?

(A) Food and Drug Administration

(B) government

(C) the general public

(D) cosmetics industry

2. Which of the following is true from the passage?

(A) We can not have a product that is free of contaminants.

(B) Too much preservative in a product is very good.

(C) Cosmetics must be preserved adequately to kill the microorganisms that are introduced by the consumers themselves.

3. As a formulator, one must (　　).

(A) not to find the right combination of preservatives

(B) do more research to find the right combination of preservatives to kill any microorganisms.

(C) find contaminants in our products.

B. Translate the following into English.

1. 防腐剂　2. 食品和药物管理局　3. 污染

4. 微生物　5. 微生物学　6. 配方设计师

7. 微生物挑战试验

C. Reading comprehension. After reading a passage, choose the best answer to each question.

With the popularity of all natural cosmetic products, it becomes more difficult as many of the current preservatives on the market today are synthetic. So how do we preserve a natural product? There are some natural preservatives available today but most are not what we consider "broad spectrum", meant to kill a wide variety of microorganisms (bacteria, mold, yeast, fungus, etc.). They are, however, very active at pH 5 to 9 and at 45 to 50℃ temperatures, the range where microbes like to grow. They are usually used in conjunction with other preservatives to kill off the microorganisms that may commonly find their way into cosmetic products. This does make the formulator, s job quite challenging (no pun intended).

The major issue is that with decorative cosmetics, there is the extra phase—the color phase—which must be taken into consideration when preserving the product. Different ingredients must be used in conjunction with the colorants to stabilize the system. These new systems now become more difficult to preserve, and conventional preservation systems may not be sufficient. Then the package that is to be used the deliver the color has to be taken into consideration. Will it be a source of contamination itself?

We are very lucky though; many common, natural ingredients today have their own anti-microbial activity. They are very easy to obtain and work very well in various cosmetic applications. And they look good on the ingredient label as well. In a future technical article, we will list some of these natural "preservative" ingredients and how much you can use to effectively obtain a sufficient kill in your product.

New Words and Phases

synthetic [sin'θetik] *adj*. 合成的，人造的
broad spectrum [ˌbrɔːd 'spektrəm] *adj*. 广谱的；用途广泛的
mold [məʊld] *n*. 霉，霉菌
yeast [jiːst] *n*. 酵母
fungus ['fʌŋgəs] *n*. 真菌，霉菌；菌类
conjunction [kən'dʒʌŋkʃn] *n*. 结合
colorant ['kʌlərənt] *n*. 着色剂
package ['pækidʒ] *n*. 包，包裹

1. Why does the natural preservative available today make the formulator' s job quite challenging?

(A) They are very active at pH5 to 9 and at 45 to 50℃ temperatures, the range where microbes like to grow.

(B) They can kill off the microorganisms that may commonly find their way into cosmetic products by themselves.

(C) They are what we consider "broad spectrum", meant to kill a wide variety of microorganisms.

(D) We can preserve a natural product by the natural preservative available today.

2. Which must be taken into consideration when conventional preservation systems may not be sufficient?

(A) the extra phase　　(B) the color phase

(C) the package　　(D) all of the above

3. Which of the following is wrong from the passage?

(A) Natural ingredients today have their own anti-microbial activity.

(B) Natural ingredients today are very easy to obtain and work very well in various cosmetic applications.

(C) Natural ingredients today look good on the ingredient label as well.

(D) We have listed some of the natural "preservative” ingredients and how much you can use to effectively obtain a sufficient kill in your product.

Reading Material

Preservatives in Cosmetics

When surfing around on the web, you will be supplied with literally thousands of pages of information dealing with preservatives, but with such a variety of conflicting ideas and data that people simply do not know who and what to believe.

To start placing this in perspective:

The most important thing is that any cosmetic products must be safe for use.

Any product manufactured without an ingredient to prevent and control microbial growth, will start to go off and may even start growing potentially pathogenic organisms.

To control microbial growth and to stabilize any cosmetic product, some form of preservative

化妆品中的防腐剂

当你在网上浏览的时候，你会看到成千上万页的关于防腐剂的信息。但是面对各种各样的相互矛盾的想法和数据，人们根本不知道该相信谁和什么。

客观地来看这个问题：

最重要的是，任何化妆品都必须安全使用。

任何没有预防和控制微生物生长成分的产品，都会开始变质，甚至可能开始生长潜在的致病微生物。

为了控制微生物的生长和稳定任何化妆品，都需要使用某种形式的防腐剂。

needs to be used.

The downside of preservatives also has some compelling arguments. Some ingredients can cause allergies in susceptible people, including dermatitis and other side effects.

防腐剂的缺点也有一些令人信服的论据。有些成分可能导致易感人群过敏，包括皮炎和其他副作用。

The most important word to keep in mind when discussing preservatives, is the word "balance". You need to include enough preservative to control microbial growth, yet not too much so as to cause allergies. dermatitis or any side effects.

在讨论防腐剂时，要记住的最重要的一个词是“平衡”。你需要加入足够的防腐剂来控制微生物的生长，但不要太多，以免引起过敏、皮炎或任何副作用。

When you wish to achieve this balance it takes both time and money, as you need to do microbial and stability tests on your products.

当你希望达到这种平衡时，需要花费时间和金钱，因为你需要对你的产品进行微生物和稳定性测试。

Should a manufacturer decide not to spend the time or money on testing and refining a formula, a stable product can still be made, by including high amounts of preservatives. This will stop microbial growth, but can cause all the negative effects: dermatitis, allergies and other side effects.

如果制造商决定不花时间或金钱在测试和精炼配方上，通过添加大量防腐剂，仍然可以生产出稳定的产品。这将阻止微生物的生长，但可能导致所有的负面影响：皮炎、过敏和其他副作用。

Glossary

A

absorb [əbˈsɔːb] *vt*. 吸收

accurate [ˈækjurit] *adj*. 正确的，精确的

achieve [əˈtʃiːv] *vt*. 完成，达到

acid [ˈæsid] *n*. 酸

administration [ədˌminiˈstreiʃn] *n*. 管理，行政；管理部门，行政部门

adsorption [ædˈsɔːpʃən] *n*. 吸附

advantage [ədˈvaːntidʒ] *n*. 优势，有利条件

aeroplane [ˈɛərəplein] *n*. 飞机

aesthetic [iːsˈθetik] *adj*. 美的；美学的

alcohol [ˈælkəhɔl] *n*. 酒精，酒

algae [ˈældʒiː] *n*. 藻类；海藻

ammonia [ˈæməunjə] *n*. 氨，氨水

anabolic [ˌænəˈbɒlik] *adj*. 合成代谢的；同化的

analyte [ˈænəlait] *n*.（被）分析物

analytical [ˌænəˈlitikəl] *adj*. 分析的，解析的

anion [ˈænaiən] *n*. 阴离子

antibiotics [ˌæntibaiˈɒtiks] *n*. 抗生素；抗生学

antiquity [æn'tikwiti] *n*. 古代，古老，古代的遗物
antiseptic [ˌænti'septik] *adj*. 防腐的，抗菌的
argon ['aːgɔn] *n*. 氩
aromatherapy [əˌrəʊmə'θerəpi] *n*. 芳香疗法
assimilate [ə'siməleit] *vi*. 吸收；同化
assume [ə'sjuːm] *vt*. 假定，设想
atmosphere ['ætməsfiə] *n*. 大气，空气
atom ['ætəm] *n*. 原子
attract [ə'trækt] *vt*. 吸引，有吸引力
attributable [ə'tribjʊtəbl] *adj*. 可归于……的
auxiliary [ɔːg'ziljəri] *adj*. 辅助的

B

bacteria [bæk'tiəriə] *n*. 细菌
balance ['bæləns] *n*. 平衡
base [beis] *n*. 底部，基础 *vt*. 以……作基础
base [beis] *n*. 碱
basin ['beisn] *n*. 盆，盆地，水池
behave [bi'heiv] *v*. 举动，举止，行为
benzene ['benziːn] *n*. 苯
beverage ['bevəridʒ] *n*. 饮料
biomass ['baiəʊmæs] *n*. 生物量

bioreactor [ˈbaiori，æktər] n.[细胞] 生物反应器

biotechnology [ˌbaiətekˈnɒlədʒi] n. 生物技术，生物工艺

bitter [ˈbitə] adj. 苦的

blubber [ˈblʌbə] n. 鲸脂

boil [bɔil] v. 煮沸

bottom [ˈbɔtəm] n. 底，底部

break [breik] v. 打破

brick [brik] n. 砖，砖块

broad spectrum [ˌbrɔːd ˈspektrəm] adj. 广谱的；用途广泛的

builder [ˈbildə] n. 增洁剂

buret [bjuəˈret] n. 滴定管，玻璃量管

butadiene [ˌbjuːtəˈdaiiːn] n. 丁二烯

butane [ˈbjuːtein] n. 丁烷

C

calcium [ˈkælsiəm] n. 钙

calculate [ˈkælkjuleit] v. 计算

calm [kaːm] v.（使）平静，（使）镇定，平息

carbon [ˈkaːbən] n. 碳

carriage [ˈkæridʒ] n. 马车，客车

carrier [ˈkæriə] *n*. 运送者，携带者，载体

case [keis] *n*. 案例，情形

catabolic [ˌkætəˈbɒlik] *adj*. 分解代谢的；异化的

catalyse [ˈkætəlaiz] *vt*. 催化；促成

catalysis [kəˈtælisis] *n*. 催化作用

catalyst [ˈkætəlist] *n*. 催化剂

catalyze [ˈkætəlaiz] *vt*. 催化

category [ˈkætigəri] *n*. 种类

cation [ˈkætaiən] *n*. 阳离子

cationic [ˈkætaiənik] adj. 阳离子的

cause [kɔːz] *n*. 原因 *vt*. 引起

cease [siːs] *v*. 停止

cellulite [ˈseljʊlait] *n*. 脂肪团

characteristic [ˌkæriktəˈristik] *n*. 特性，特征

charge [tʃaːdʒ] *n*. 电荷

chemistry [ˈkemistri] *n*. 化学

chlorine [ˈklɔːriːn] *n*. 氯，氯气

chromatography [ˌkrəuməˈtɔgrəfi] *n*. 色谱法

claims [kleimz] *n*. 要求，请求权

classification [ˌklæsifiˈkeiʃən] *n*. 分类，分级

classify [ˈklæsifai] *vt*. 分类，分等

climate [ˈklaimit] *n*. 气候

clump [klʌmp] *v*. 使成团或成块，使结合
coarse [kɔːs] *adj*. 粗糙的
coating ['kəutiŋ] *n*. 涂料
collector [kə'lektə] *n*. 收集器
colloid ['kɔlɔid] *n*. 胶体
colorant ['kʌlərənt] *n*. 着色剂
combination [ˌkɔmbi'neiʃən] *n*. 结合，联合，化合
combine [kəm'bain] *v*.（使）联合，（使）结合
combustion [kəm'bʌstʃən] *n*. 燃烧
comparatively [kəm'pærətivli] *adv*. 比较地，相当地
component [kəm'pəunənt] *n*. 成分
composition [kɔmpə'ziʃən] *n*. 成分，组成
compound ['kɔmpaund] *n*. 化合物
compress [kəm'pres] *vt*. 压缩
comprise [kəm'praiz] *v*. 包含，由……组成
concentration [ˌkɔnsen'treiʃən] *n*. 浓度
conclude [kən'kluːd] *vt*. 推断，断定
condenser [kən'densə] *n*. 冷凝器
condition [kən'diʃən] *n*. 条件，环境
conduct ['kɔndʌkt] *v*. 传导
conduction [kən'dʌkʃən] *n*. 传导
confident ['kɔnfidənt] *adj*. 自信的，确信的

conjunction [kən'dʒʌŋkʃn] *n*. 结合
constant ['kɔnstənt] *n*. 常数
consume [kən'sjuːm] *vt*. 消耗
consumption [kən'sʌmpʃən] *n*. 消费
contain [kən'tein] *vt*. 包含
contamination [kənˌtæmi'neiʃn] *n*. 污染
contract ['kɔntrækt] *v*. 收缩
contribute [kən'tribjuːt] *v*. 贡献，促进
convection [kən'vekʃən] *n*. 对流
conversely ['kɔnvɜːsli] *adv*. 相反地
convert [kən'vəːt] *vt*. 使转变，转换
corridor ['kɔridɔː] *n*. 走廊
corrode [kə'rəud] *v*. 使腐蚀，侵蚀
cosmetic [kɒz'metik] *adj*. 化妆用的，美容的；*n*. 化妆品，美容品
countless ['kaʊtlis] *adj*. 无数的，数不尽的
creature ['kriːtʃə] *n*. 动物
crowd [kraud] *n*. 人群，群众
crude [kruːd] *adj*. 天然的，未加工的
crush [krʌʃ] *vt*. 压碎，碾碎
crystallization ['kristəlai'zeiʃən] *n*. 结晶
cubic meter 立方米

culture [ˈkʌltʃər] *n.* 培养，培养物

D

dam [dæm] *n.* 水坝

dandruff [ˈdændrʌf] *n.* 头皮屑

decay [diˈkei] *vi.* 腐朽，腐烂

decompose [ˌdiːkəmˈpəuz] *v.* 分解

define [diˈfain] *vt.* 定义，详细说明

dense [dens] *adj.* 密集的，浓厚的

density [ˈdensiti] *n.* 密度

deposit [diˈpɔzit] *n.* 堆积物，沉淀物 *v.* 存放，堆积

derive [diˈraiv] *vi.* 起源

despite [disˈpait] *prep.* 不管，尽管，不论

destroy [disˈtrɔi] *vt.* 破坏，毁坏，消灭

detect [diˈtekt] *vt.* 探测，测定

detector [diˈtektə] *n.* 探测器

detergent [diˈtəːdʒənt] *n.* 清洁剂，去垢剂

determination [ditəːmiˈneiʃən] *n.* 测量，测定

diagnosis [ˌdaɪəgˈnəʊsis] *n.* 诊断，判断

diesel [ˈdiːzəl] *n.* 柴油机

differ [ˈdifə] *vi.* 不一致，不同

diffuse [diˈfjuːz] *v.* 散播，传播，散开

digest [di'dʒest] *v*. 消化

digestion [dai'dʒestʃən] *n*. 消化；消化能力

dispersion [dis'pəːʃən] *n*. 散布，驱散，传播

disperse [di'spɜːs] *v*. 分散，散布

dissolve [di'zɔlv] *v*. 溶解

distill [di'stil] *vt*. 提取；蒸馏

divide [di'vaid] *v*. 分，划分，分开

double ['dʌbl] *vt*. 使加倍

dust [dʌst] *n*. 灰尘，尘土

E

effect [i'fekt] *n*. 结果，效果

effectiveness [i'fektivnis] *n*. 效力，有效性

elastomer [i'læstəmə] *n*. 弹性体，人造橡胶

electricity [ilek'trisiti] *n*. 电流，电，电学

electron [i'lektrɔn] *n*. 电子

element ['elimənt] *n*. 元素，要素

elevation [ˌeli'veiʃən] *n*. 上升，提高

emerge [i'məːdʒ] *vi*. 显现，形成

emulsion [i'mʌlʃn] *n*. 乳剂；乳状液

emit [i'mit] *vt*. 发出，散发

enhance [in'hɑːns] *v*. 提高

enzyme ['enzaim] *n*. 酶

equilibrium [ˌiːkwi'libriəm] *n*. 平衡

equipment [i'kwipmənt] *n*. 装备，设备

ester ['estə] *n*. 酯

estimate ['estimeit] *v*. 估计，估价，评估

evaporation [iˌvæpə'reiʃən] *n*. 蒸发

exceed [ik'siːd] *vt*. 超越，胜过

excess [ik'ses] *adj*. 超过的，额外的

existence [ig'zistəns] *n*. 存在，实在

expand [iks'pænd] *vt*. 使膨胀，扩张

explain [iks'plein] *v*. 解释，说明

explanation [ˌekspləˈneiʃən] *n*. 解释，解说，说明

expose [iks'pəuz] *v*. 使暴露，揭露

exposure [iks'pəuʒə] *n*. 暴露，揭露

extend [iks'tend] *v*. 扩充，延伸

F

false [fɔːls] *adj*. 错误的

favor ['feivə] *vt*. 促成，支持

feed [fiːd] *v*. 进料，加料

feedstuff ['fiːdstʌf] *n*. 饲料，原料

feedstock ['fiːdstɔk] *n*. 给料

fermentation [ˌfəːmenˈteiʃn] *n*. 发酵

fertilizer [ˈfəːtiˌlaizə] *n*. 肥料

fiber [ˈfaibə] *n*. 纤维

figure [ˈfigə] *n*. 图形

filtrate [ˈfiltreit] *n*. 滤出液

filtration [filˈtreiʃən] *n*. 过滤

flour [ˈflauə] *n*. 面粉

fluid [ˈfluːid] *n*. 流体，流动性

force *n*. 力，力量

form [fɔːm] *n*. 形状 *v*. 形成，构成

formulator [ˈfɔːmjuleitə] *n*. 配方设计师

fossil [ˈfɔsl] *adj*. 化石的

friction [ˈfrikʃən] *n*. 摩擦，摩擦力

fuel [fjuəl] *n*. 燃料

function [ˈfʌŋkʃən] *n*. 官能，功能

fungi [ˈfʌŋgiː] *n*. 真菌，霉菌；菌类

fungus [ˈfʌŋgəs] *n*. 真菌，霉菌；菌类

G

gasoline [ˈgæsəliːn] *n*. 汽油

generate [ˈdʒenəˌreit] *vt*. 产生，发生

generation [ˌdʒenəˈreiʃən] *n*. 产生

give off 发出（蒸汽、光等）

govern [ˈgʌvən] *v*. 统治，支配，管理

grain [grein] *n*. 细粒，颗粒

grease [griːs] *n*. 油脂

grind [graind] *v*. 磨碎，碾碎

H

handle [ˈhændl] *vt*. 处理，操作

harm [haːm] *vt*. 伤害，损害

hold [həuld] *v*. 保持

hormone [ˈhɔːməʊn] *n*. [生理] 激素，荷尔蒙

humidity [hjuːˈmiditi] *n*. 湿气，潮湿

hydrate [ˈhaidreit] *n*. 氢氧化物，*v*. 与水化合

hydrocarbon [ˈhaidrəuˈkaːbən] *n*. 烃，碳氢化合物

hydrogen [ˈhaidrəudʒən] *n*. 氢，氢气

hydrolysis [haiˈdrɔlisis] *n*. 水解

hydrophilic [ˌhaidrəuˈfilik] *adj*. 亲水的，吸水的

hydrophobic [ˌhaidrəʊˈfəʊbik] *adj*. 不易被水沾湿的，疏水的

hydroxide [haiˈdrɔksaid] *n*. 氢氧化物

hypochlorite [ˌhaipəuˈklɔːrait] *n*. 次氯酸盐

I

identify [aiˈdentifai] *vt*. 识别，鉴别

illumination [iˌljuːmiˈneiʃən] *n*. 照明，阐明，启发
imagine [iˈmædʒin] *vt*. 想象，设想
impossibility [imˌpɔsəˈbiləti] *n*. 不可能之事，不可能
include [inˈkluːd] *vt*. 包括，包含
increase [inˈkriːs] *n*. 增加 *v*. 增加
indicator [ˈindikeitə] *n*. 指示剂
inert [iˈnəːt] *adj*. 惰性的
ingredient [inˈgriːdiənt] *n*. 成分
inflammation [ˌinfləˈmeiʃn] *n*. 炎症
initial [iˈniʃəl] *adj*. 最初的，初始的
injection [inˈdʒekʃən] *n*. 注射
inorganic [ˌinɔːˈgænik] *adj*. 无机的
insoluble [inˈsɔljubl] *adj*. 不能溶解的
interact [ˌintərˈækt] *vi*. 互相作用，互相影响
interferon [ˌintəˈfiərɒn] *n*. [生化] 干扰素
internal combustion engine 内燃机
investment [inˈvestmənt] *n*. 投资
invigorate [inˈvigəreit] *vt*. 鼓舞；使精力充沛
ion [ˈaiən] *n*. 离子
iron [ˈaiən] *n*. 铁

J

juice [dʒuːs] *n*.（水果）汁，液

K

key [kiː] *n*. 关键

kinetic [kaiˈnetik] *adj*. 运动的，动的

krypton [ˈkriptɔn] *n*. 氪

L

labeling [ˈleibliŋ] *n*. 标签；标记

lattice [ˈlætis] *n*. 格子

lead [liːd] *n*. 铅

lecture [ˈlektʃə] *n*. 演讲

legal [ˈliːgl] *adj*. 法律的；合法的；法定的

lipstick [ˈlipstik] *n*. 口红，唇膏

liter [ˈliːtə] *n*. 公升

living [ˈliviŋ] *adj*. 活的

load [ləud] *n*. 负荷，负载 *v*. 装载

locomotive [ˌləukəˈməutiv] *n*. 机车，火车头

lower [ˈləuə] *v*. 降低，减弱

lubricate [ˈluːbrikeit] *vt*. 润滑

M

macroemulsion [ˈmækrəʊˈmʌlʃn] *n*. 粗乳液

magnesium [mæg'ni:zjəm] *n*. 镁

magnet ['mægnit] *n*. 磁铁

mandate ['mændeit] *v*. 强制执行，颁布

manufacture [ˌmænju'fæktʃə] *vt*. 制造，加工

marine [mə'ri:n] *adj*. 海的，航海的

mark [ma:k] *n*. 标志，分数，记号 *vt*. 做标记，打分数

mass [mæs] *n*. 块，质量

measure ['meʒə] *v*. 测量

measure ['meʒə] *vt*. 测量

medium ['mi:djəm] *n*. 媒介

melt [melt] *v*.（使）融化，（使）熔化

mercury ['mə:kjuri] *n*. 水银，汞

metabolic [ˌmetə'bɒlik] *adj*. 新陈代谢的

methane ['meθein] *n*. 甲烷，沼气

microbiology [ˌmaɪkrəʊbai'ɒlədʒi] *n*. 微生物学

microemulsion ['maɪkrəʊ'mʌlʃn] *n*. 微乳液

microorganism [maikrəʊ'ɔ:gənɪzəm] *n*.［微］微生物

microscope ['maikrəskəʊp] *n*. 显微镜

microscopic [maikrə'skɔpik] *adj*. 用显微镜可见的

mineral ['minərəl] *n*. 矿物，矿石

minimize ['minimaiz] *v*. 使减少到最低限度；贬低

minireactor [ˌminiriˈseʃən] *n.* 小反应器
mitigation [ˌmitiˈgeiʃn] *n.* 减轻；缓和；平静
mix [miks] *v.* 使混合，混合
mixture [ˈmikstʃə] *n.* 混合，混合物
moist [mɔist] *adj.* 潮湿的 *n.* 潮湿
moisture [ˈmɔistʃə] *n.* 潮湿，湿气
mold [məʊld] *n.* 霉，霉菌
molecule [ˈmɔlikjuːl] *n.* 分子
monitor [ˈmɔnitə] *vt.* 监控
monomer [ˈmɔnəmə] *n.* 单体
motion [ˈməuʃən] *n.* 运动，动作
movement [ˈmuːvmənt] *n.* 运动，动作

N

natural [ˈnætʃərəl] *adj.* 自然的，天生的
negative [ˈnegətiv] *adj.* 负的，阴性的
neon [ˈniːən] *n.* 氖
net [net] *adj.* 净余的
neutral [ˈnjuːtrəl] *adj.* 中性的
neutralize [ˈnjuːtrəlaiz] *v.* 中和
nitrogen [ˈnaitrədʒən] *n.* 氮，氮气
note [nəut] *vt.* 注意

O

object [ˈɔbdʒikt] *n*. 物体

obtain [əbˈtein] *vt*. 获得，得到

occur [əˈkəː] *vi*. 发生，出现

oppose [əˈpəuz] *vt*. 反对，使对立

optimization [ɒptimaiˈzeiʃən] *n*. 优化；最优化

ordered [ˈɔːdəid] *adj*. 规则的，整齐的

organic [ɔːˈgænik] *adj*. 器官的，有机的

overland [ˈəuvəlænd] *adj*. 陆路的，陆上的

owe [əu] *v*. 把……归功于，欠

oxygen [ˈɔksidʒən] *n*. 氧，氧气

oxygenate [ˈɒksidʒəneit] *v*.（使）富氧；氧化

P

package [ˈpækidʒ] *n*. 包，包裹

paint [peint] *n*. 油漆；颜料

paraffin-oil [ˈpærəfin-ɔil] *n*. 石蜡油

particle [ˈpaːtikl] *n*. 粒子，微粒

perfectly *adv*. 很，完全，完美地

perfume [ˈpəːfjuːm] *n*. 香味，芳香，香水

peroxide [pəˈrɔksaid] *n*. 过氧化物

petrol [ˈpetrəl] *n.* 汽油

petroleum [piˈtrəuliəm] *n.* 石油

pharmaceutical [ˌfaːməˈsjuːtikəl] *adj.* 药物的

phase [feiz] *n.* 相

physics [ˈfiziks] *n.* 物理学

pickle [ˈpikl] *n.* 泡菜；盐卤；腌制食品

pipeline [ˈpaipˌlain] *n.* 管道，管线

plastic [ˈplæstik] *n.* 塑胶

platinum [ˈplætinəm] *n.* 白金，铂

pollutant [pəˈluːtənt] *n.* 污染物

polish [ˈpɒliʃ] *n.* 磨光，擦亮；擦亮剂

polybutadiene [ˌpɔliˌbjuːtəˈdaiiːn] *n.* 聚丁二烯

polymer [ˈpɔlimə] *n.* 聚合物

porous [ˈpɔːrəs] *adj.* 多孔的

positive [ˈpɔzətiv] *adj.* 正的，阳性的

potential [pəˈtenʃəl] *adj.* 潜在的，可能的

pour [pɔːr] *vi.* 灌，倒；倾泻

precipitate [priˈsipiteit] *v.* 沉淀

precipitation [priˌsipiˈteiʃən] *n.* 沉淀

presence [ˈprezns] *n.* 存在

preservative [priˈzɜːvətiv] *n.* 防腐剂，保护剂

pressure [ˈpreʃər] *n.* 压，压力

primary [ˈpraiməri] *adj*. 主要的，初步的，初级的

principle [ˈprinsəpl] *v*. 原则，原理

proceed [prəˈsiːd] *vi*. 进行，继续

profit [ˈprɔfit] *n*. 利润，益处，得益

promote [prəˈməut] *vt*. 促进，提升

protein [ˈprəutiːn] *n*. 蛋白质

proton [ˈprəutɔn] *n*. 质子

provide [prəˈvaid] *v*. 供应，供给

psychological [ˌsaikəˈlɒdʒikl] *adj*. 心理的，精神的

pump [pʌmp] *n*. 泵

purification [ˌpjuərifiˈkeiʃən] *n*. 净化，纯化

pyrometer [ˌpaiəˈrɔmitə] *n*. 高温计

Q

quantity [ˈkwɔntiti] *n*. 量，数量

quart [kaːt] *n*. 夸脱（容量单位）

R

radiation [ˌreidiˈeiʃən] *n*. 发散，发光，辐射

range [reindʒ] *v*. 在……范围

reactant [riːˈæktənt] *n*. 反应物

reactor [ri(ː)ˈæktə] *n*. 反应器

readily [ˈredili] *adv*. 容易地

realize [ˈriəlaiz] *vt*. 认识到，了解

reasonable [ˈriːznəbl] *adj*. 合理的，有道理的

reciprocate [riˈsiprəkeit] *v*. 往复

recovery [riˈkʌvəri] *n*. 恢复

redissolve [ˈriːdiˈzəlv] *v*. 再溶解

refine [riˈfain] *vt*. 精炼，精制

refrigerator [riˈfridʒəreitə] *n*. 电冰箱

release [riˈliːs] *vt*. 释放

remain [riˈmein] *vi*. 保持，剩余

removal [riˈmuːvəl] *n*. 去除

remove [riˈmuːv] *v*. 去除

residue [ˈrezidjuː] *n*. 滤渣

resistance [riˈzistəns] *n*. 抵抗，抵抗力

restore [riˈstɔː] *vt*. 恢复；修复；归还

retain [riˈtein] *vt*. 保持，保留

retard [riˈtaːd] *vt*. 延迟，阻碍

reversible [riˈvəːsəbl] *adj*. 可逆的

revitalize [riːˈvaitəlaiz] *vt*. 使……复活；使……复兴

rotary [ˈrəutəri] *adj*. 旋转的

rust [rʌst] *n*. 铁锈 *vt*.（使）生锈

S

sample [ˈsæmpl] *n.* 样品

saturate [ˈsætʃəreit] *v.* 饱和

saturation point *n.* 饱和点

sausage [ˈsɒsidʒ] *n.* 香肠；腊肠

screen [skriːn] *v.* 分筛

secondary [ˈsekəndəri] *adj.* 二级的，中级

semitransparent [ˌsemitrænsˈpærənt] *adj.* 半透明的

sensitivity [ˈsensiˈtiviti] *n.* 敏感，灵敏性

separate [ˈsepəreit] *adj.* 分开的，分离的 *v.* 分开，隔离

serve [səːv] *v.* 服务，供应

settle [ˈsetl] *v.* 澄清，沉淀

shampoo [ʃæmˈpuː] *n.* 洗发；洗发精

shape [ʃeip] *n.* 外形，形状，形态

silver chloride 氯化银

simplicity [simˈplisiti] *n.* 简单，简易

slippery [ˈslipəri] *adj.* 滑的，光滑的

soak [səuk] *v.* 浸，泡，浸透

sodium [ˈsəudiəm] *n.* 钠

sodium sulfate *n.* 硫酸钠

soil [sɔil] *n*. 污垢

solar ['səulə] *adj*. 太阳的，日光的

solubility [ˌsɔlju'biliti] *n*. 溶解度，溶解性

soluble ['sɔljubl] *adj*. 可溶的，可溶解的

solute ['sɔljuːt] *n*. 溶质

solution [sə'ljuːʃən] *n*. 溶液

sour ['sauə] *adj*. 酸的，酸味的

space [speis] *n*. 空间

specialize ['speʃəlaiz] *vi*. 专攻，专门研究

spill [spil] *v*. 溢出，溅出

spoon [spuːn] *n*. 匙，勺子

spray [sprei] *n*. 喷雾，喷雾剂；*vt*. 喷射

spread [spred] *v*. 伸展，展开

sprinkle ['spriŋkl] *vi*. 撒，洒；用……点缀

stage [steidʒ] *n*. 阶段，时期

stationery ['steiʃənəri] *adj*. 固定的

steam [stiːm] *n*. 蒸汽，水汽

sterilized ['sterəlaizd] *adj*. 无菌的；已消过毒的

stimulate ['stimjuleit] *vt*. 刺激；鼓舞，激励

stir [stəː] *vt*. 摇动，搅和

stomach ['stʌmək] *n*. 胃，胃部

storage ['stɔridʒ] *n*. 贮藏，存储

strip [strip] *n.* 条，带

strip off *v.* 剥落

subcellular [sʌbˈseljʊlə (r)] *adj.* 亚细胞的

sugar [ˈʃugə] *n.* 糖

sulfur [ˈsʌlfə] *n.* 硫，硫黄

superior [sjuːˈpiəriə] *adj.* 较高的，上好的

surfactant [səːˈfæktənt] *n.* 表面活性剂

surplus [ˈsəːpləs] *adj.* 过剩的，剩余的

suspend [səsˈpend] *vt.* 悬挂，悬浮

suspension [səsˈpenʃən] *n.* 悬浮，悬浮液

synthetic [sinˈθetic] *adj.* 合成的，人造的

T

tank [tæŋk] *n.* 槽，箱，罐，釜

term [təːm] *n.* 术语

terminus [ˈtəːminəs] *n.* 终点

therapeutic [θerəˈpjuːtik] *adj.* 治疗的；治疗学的

thermometer [θəˈmɔmitər] *n.* 温度计，体温计

thermoplastic [ˌθəːməˈplæstik] *adj.* 热塑性的

thermosetting [ˌθəːməuˈsetiŋ] *adj.* 热固性的

titrant [ˈtaitrənt] *n.* 滴定剂

titrate [ˈtaitreit] *v.* 滴定，用滴定法测量

titration [tai'treiʃən] *n.* 滴定

tonnage ['tʌnidʒ] *n.* 吨位

toss [tɔs] *v.* 投，掷

transfer [træns'fəː] *n.* 迁移，移动 *v.* 转移，传递

transparent [træns'pærən] *adj.* 透明的，清澈的

transport [træns'pɔːt] *vt.* 传送，运输

troposphere ['trɔpəusfiə] *n.* 对流层

U

undergo [ˌʌndə'gəu] *vt.* 经历，经过

unit ['juːnit] *n.* 个体，单位

unwelcome [ʌn'welkəm] *adj.* 不受欢迎的

upmost ['ʌpməust] *adj.* 最高的，最上的

V

vaccine ['væksiːn] *n.* 疫苗

vacuum ['vækjuəm] *n.* 真空

vapor ['veipə] *n.* 水蒸气

vaporize ['veipəraiz] *v.*（使）蒸发

vegetable ['vedʒitəbl] *n.* 蔬菜，植物

velocity [vi'lɔsiti] *n.* 速度，速率

vinegar ['vinigə] *n.* 醋

violate [ˈvaiəleit] *vt*. 违反；侵犯，妨碍；亵渎

visible [ˈvizəbl] *adj*. 看得见的

volume [ˈvɔljuːm] *n*. 体积

W

warfare [ˈwɔːfɛə] *n*. 战争

warship [ˈwɔːʃip] *n*. 军舰，战船

well [wel] *n*. 井

whale [weil] *n*. 鲸

X

xenon [ˈzenɔn] *n*. 氙

Y

yeast [jiːst] *n*. 酵母

yield [jiːld] *n*. 产量，收益

yogurt [ˈjɒgət] *n*. 酸奶酪，酸乳

References

[1] 李居参.工业分析专业英语.3版.北京：化学工业出版社，2018.

[2] 符德学.化学化工专业英语.3版.北京：化学工业出版社，2020.

[3] 丁慧，马晓燕，李成林.应用化学化工专业英语.哈尔滨：哈尔滨工程大学出版社，2007.

[4] 文杰，麟伟.牛津图解中学化学.一毅，译.上海：上海教育出版社，2001.

[5] 约翰逊.国家地理科学探索丛书：酸还是碱?北京：外语教育与研究出版社，2005.

[6] 特斯特，等.能源.王永，译.北京：外语教育与研究出版社，2005.

[7] 邬行彦，储炬，宫衡.生物工程生物技术专业英语.北京：化学工业出版社，2019.